PREDICTIVE CONTROL OF POWER CONVERTERS AND ELECTRICAL DRIVES

PREDICTIVE CONTROL OF POWER CONVERTERS AND ELECTRICAL DRIVES

Jose Rodriguez and Patricio Cortes
Universidad Tecnica Federico Santa Maria, Valparaiso, Chile

A John Wiley & Sons, Ltd., Publication

This edition first published 2012

Registered office
John Wiley & Sons Ltd, The Atrium, Southern Gate, Chichester, West Sussex, PO19 8SQ, United Kingdom

For details of our global editorial offices, for customer services and for information about how to apply for permission to reuse the copyright material in this book please see our website at www.wiley.com.

Library of Congress Cataloging-in-Publication Data

Rodriguez Perez, Jose.
Predictive control of power converters and electrical drives / Rodriguez Perez, Jose, Patricio Cortes Estay.
p. cm.
Includes bibliographical references and index.
ISBN 978-1-119-96398-1 (cloth)
1. Electric driving–Automatic control. 2. Electric current converters–Automatic control. 3. Predictive control.
I. Estay, Patricio Cortes. II. Title.
TK4058.R64 2012
621.3815'322–dc23

2011049804

A catalogue record for this book is available from the British Library.
ISBN: 9781119963981

Typeset in 10/12 Times by Laserwords Private Limited, Chennai, India

Contents

Foreword

Predictive Control of Power Converters and Electrical Drives is an essential work on modern methodology that has the potential to advance the performance of future energy processing and control systems. The main features of modern power electronic converters such as high efficiency, low size and weight, fast operation and high power densities are achieved through the use of the so-called *switch mode operation*, in which power semiconductor devices are controlled in ON/OFF fashion (operation in the active region is eliminated). This leads to different types of pulse width modulation (PWM), which is the basic energy processing technique used in power electronic systems. The PWM block not only controls but also linearizes power converters, thus it can be considered as a linear power amplifier (actuator). Therefore, power converter and drive systems classically are controlled in cascaded multi-loop systems with PI regulators.

Model-based predictive control (MPC) offers quite a different approach to energy processing, considering a power converter as a discontinuous and nonlinear actuator. In the MPC system the control action is realized in a single controller by on-line selection from all possible states, calculated in the discrete-time predictive model only as the one which minimizes the cost function. Therefore, by appropriate cost function formulation it allows larger flexibility and also achieves the optimization of several important parameters like number of switchings, switching losses, reactive power control, motor torque ripple minimization, etc. Thus, the predictive controller takes over the functions of the PWM block and cascaded multi-loop PI control of a classical system, and can offer to industry flexibility, simplicity and software-based optimal solutions where several objectives must be fulfilled at the same time. The price which is paid for the use of a predictive controller is the large number of calculations required. However, it goes well with the fast development of signal processor capacities and the evolution of industrial informatics.

In 13 chapters organized in four parts, the authors cover the basic principles of predictive control and introduce the reader in a very systematic way to the analysis and design methodology of MPC systems for power converters and AC motor drives. The book has the typical attributes of a monograph. It is well organized and easy to read. Several topics are discussed and presented in a very original way as a result of the wide research performed by the authors. The added simulation examples make the book attractive to researchers, engineering professionals, undergraduate/graduate students of electrical engineering and mechatronics faculties.

Finally, I would like to congratulate the authors for their persistence in research work on this class of control systems. I do hope that the presented work will not only perfectly fill the gap in the book market, but also trigger further study and practical implementation of predictive controllers in power electronics and AC drives.

Marian P. Kazmierkowski
Warsaw University of Technology, Poland

Preface

Although model predictive control (MPC) has been in development over some decades, its application to power electronics and drives is rather recent, due to the fast processing time required to control electrical variables.

The fast and powerful microprocessors available today have made it possible to perform a very large number of calculations at low cost. Consequently, it is now possible to apply MPC in power electronics and drives. MPC has a series of characteristics that make it very attractive: it is simple, intuitive, easy to implement, and can include nonlinearities, limitations. etc.

MPC has the potential to change dramatically how we control electrical energy using power converters.

The book is organized in four parts, covering the basic principles of power converters, drives and control, the application of MPC to power converters, the application of MPC to motor drives, and some general and practical issues on the implementation of MPC. In addition, simulation files will be available for download in the book website (http://www.wiley.com/go/rodriguez_control), allowing the reader to study and run the simulations for the examples shown in the book.

After several years of working on this topic, and considering the increasing number of journal and conference papers on it, we realized that it was becoming more and more a relevant topic. Over these years we gathered a large amount of work that was then organized as a series of lectures that were presented in several universities and later as tutorials at several international conferences. From all this material we have selected the most interesting examples and have developed some of the different chapters, trying to keep a simple and easy-to-follow explanation.

This book is intended for engineers, researchers, and students in the field of power electronics and drives who want to start exploring the use of MPC, and for people from the control theory area who want to explore new applications of this control strategy. The contents of this book can be also considered as part of graduate or undergraduate studies on advanced control for power converters and drives.

We hope that with the help of this book, more and more people will become involved in this interesting topic and new developments will appear in the forthcoming years.

Preface

Acknowledgments

The authors would like to acknowledge the support received from several people and institutions that made possible the elaboration of this book or helped in different stages of this work.

Most of the results shown in this book have been funded in part by Universidad Tecnica Federico Santa Maria, the Chilean National Fund for Scientific and Technological Development FONDECYT (under grants 1101011 and 1100404), Basal Project FB021 "Valparaiso Center for Science and Technology", Anillo Project ACT-119, and Qatar Foundation (Qatar National Research Fund grant NPRP \#4-077-2-028).

We specially thank Samir Kouro, Monina Vasquez, Rene Vargas, Hector Young, Marco Rivera, Christian Rojas, Cesar Silva, Marcelo Perez, Juan Villarroel, Juan Carlos Jarur, Sabina Torres, Mauricio Trincado, Alexis Flores, and all the students and researchers who contributed to the work that led to this book.

Finally, we acknowledge the inspiration, patience, and support of our families during the preparation of this book.

Acknowledgments

[illegible]

[illegible]

[illegible]

[illegible]

[illegible]

Part One
Introduction

1

Introduction

In the last few decades, the use of power converters and high-performance adjustable speed drives has gained an increased presence in a wide range of applications, mainly due to improved performance and higher efficiency, which lead to increased production rates. In this way, power converters and drives have become an enabling technology in most industrial sectors, with many applications in a wide variety of systems. Conversion and control of electrical energy using power electronics is a very important topic today, considering the increasing energy demands and new requirements in terms of power quality and efficiency. In order to fulfill these demands new semiconductor devices, topologies, and control schemes are being developed.

This chapter presents a basic introduction and useful references for readers who are not familiar with power converters, motor drives, and their applications. The most common applications that involve the use of power converters are presented, and a general scheme for a drive system is explained. The power converter topologies found in industry are introduced according to a simple classification. A brief introduction to control schemes for power converters, the basic concepts behind them and the digital implementation technologies used today, are discussed.

This chapter provides the necessary context, including a brief motivation for the use of predictive control, to understand the contents of this book.

1.1 Applications of Power Converters and Drives

Power converters and drives are used in diverse sectors, ranging from industrial to residential applications [1, 2]. Several application examples for different sectors are shown in Figure 1.1, where a diagram of the system configuration is shown as an example for each group, marked with $*$.

From the drive applications used in industry, pumps and fans are those that account for most of the energy consumption, with power ratings up to several megawatts. The use of adjustable speed drives can bring important benefits to these kinds of systems in terms of performance and efficiency. Many interesting applications of high-power drives can be found in the mining industry, for example, in downhill belt conveyors. A block diagram of one of these systems is shown in Figure 1.1, where three-level converters with active

Predictive Control of Power Converters and Electrical Drives, First Edition. Jose Rodriguez and Patricio Cortes.

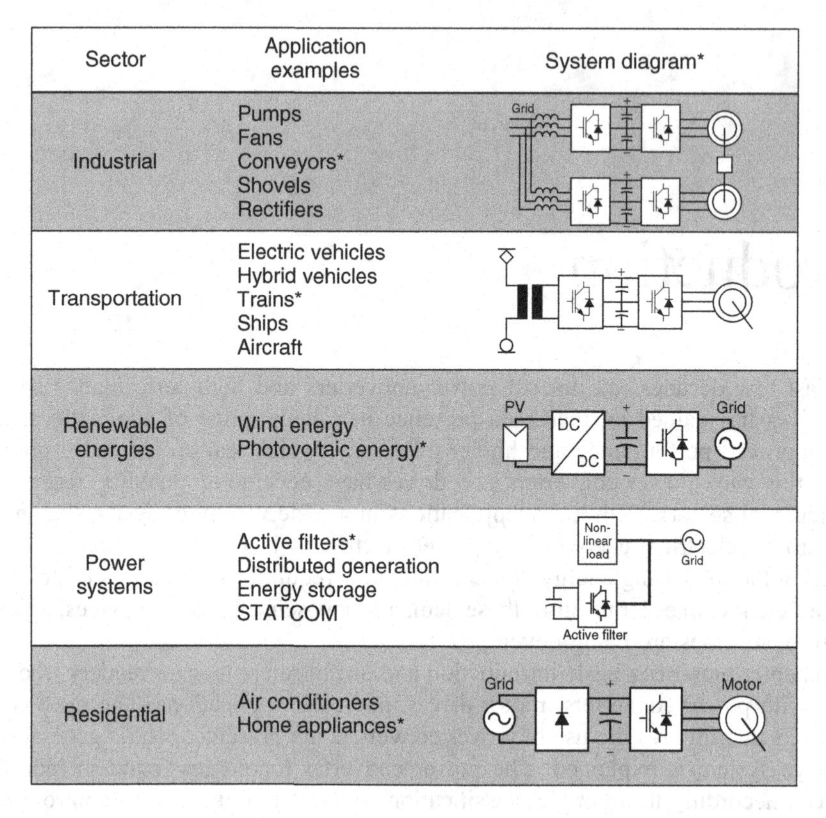

Figure 1.1 Power converter applications

front-end rectifiers are used for regenerative operation, that is, power flowing from the motors to the grid [3, 4].

Common applications of drives can be found in transportation, where electric motors are used for traction and propulsion. In electric trains, the power is transferred from the overhead lines to the motors using a power converter like the one shown in the figure. This converter generates the required voltages for controlling the torque and speed of the electric motor. High-power drives can be found in ships, where diesel engines are used as generators and the propulsion is generated by electric motors. Newer applications in transportation can be found in electric and hybrid vehicles, and in aircraft.

The use of power converters in renewable energy conversion systems has been constantly increasing in recent years, mainly due to growing energy demands and environmental concerns. Among the different renewable energy sources, photovoltaic (PV) generation systems are a very interesting example of power converter applications because it is not possible to deliver power from the PV panel to the grid without a converter. An example of a power converter for a PV system is shown in Figure 1.1, composed of a DC–DC converter for optimal operation of the panel and an inverter for injection of sinusoidal currents to the grid. The use of power converters and drives in wind generation systems allows optimization of the amount of energy extracted from the wind and compliance with the new grid regulations that impose restrictions on the power quality and performance of the system [5].

The use of power converters can help to improve the quality and stability of the grid. Some examples of power converters with applications in power systems are active filters, converters for distributed generation, energy storage systems, static VAR compensators (STATCOM), and others. A diagram of an active filter application is also shown in Figure 1.1, where the power converter generates the required currents for compensating the distorted currents generated by a nonlinear load. In this way, distortion of the grid voltage is avoided.

Low-power drives and converters offer many possibilities in residential applications. The use of adjustable speed drives can increase the efficiency of systems like air conditioners and other home appliances [1, 6].

1.2 Types of Power Converters

There are many types of power converters and drive systems, and every application requires different specifications that define the most appropriate topology and control scheme to be used. A general scheme for a drive system and a simple classification of the different types of power converters are presented next.

1.2.1 Generic Drive System

A block diagram and a picture of a real drive system are shown in Figure 1.2. The main components of the system are the line-side transformer, the rectifier, the DC link, the inverter, the electrical machine, and the control unit. Depending on the system requirements, the rectifier can be a diode rectifier or an active front-end rectifier. The DC link is composed of capacitors or inductors, depending on the topology of the inverter and rectifier, whose purpose is to store energy and decouple the operation of the inverter and rectifier. The inverter modulates the DC link voltage (or current) and generates a voltage whose fundamental component can be adjusted in amplitude, frequency, and phase, in order to control the torque and speed of the machine. The control unit samples voltage and current measurements of the most important variables and generates the gate drive signals for the power semiconductor devices.

As can be observed in Figure 1.2, the drive system requires several additional elements for proper operation, such as transformers, input and output passive filters, and a cooling system for the switching devices.

1.2.2 Classification of Power Converters

Power converters are composed of power semiconductor switches and passive components. They can be classified according to several criteria. A very simple and useful classification considers the type of conversion from input to output that the system

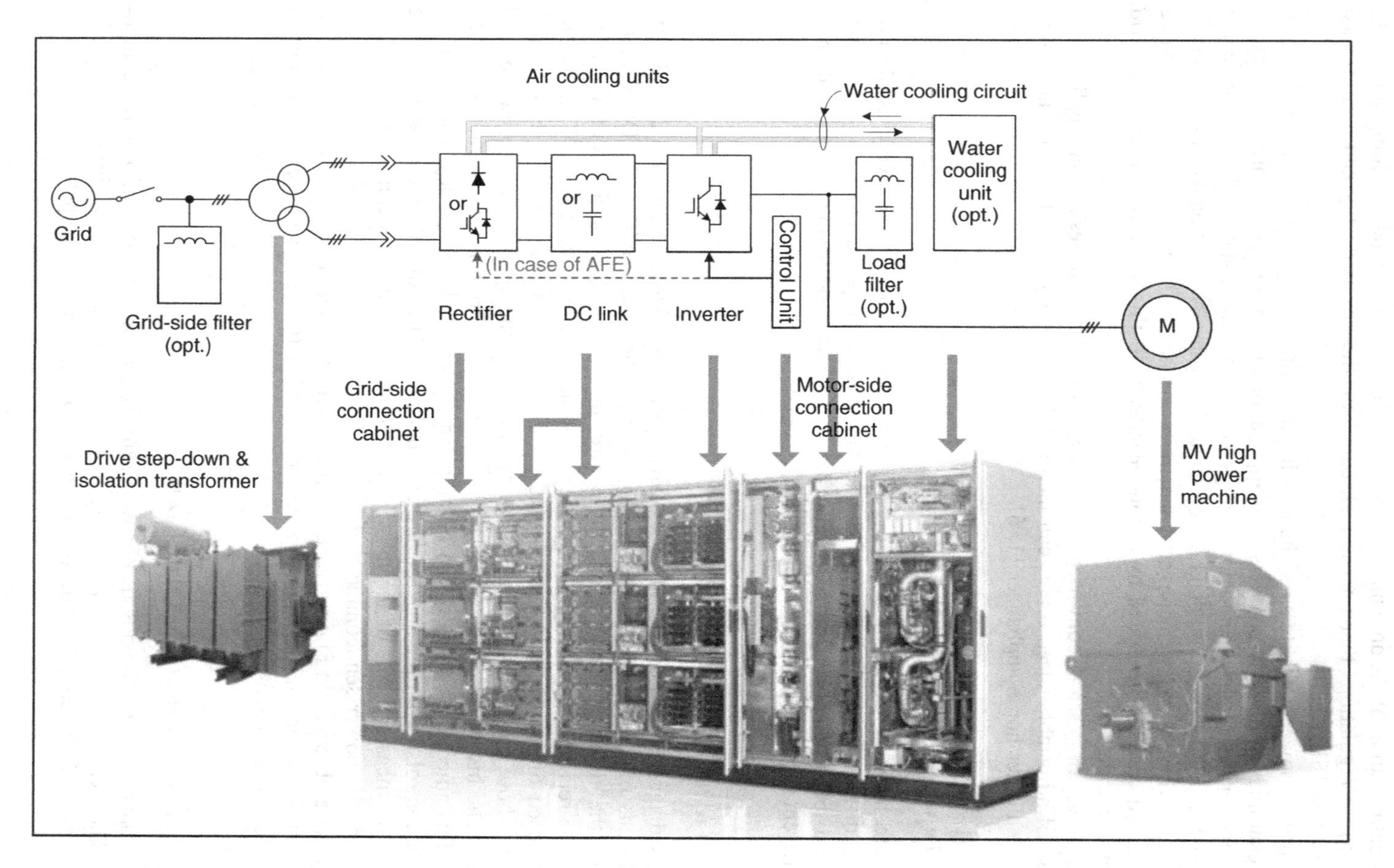

Figure 1.2 General motor drive system (AFE = Active Front-End)

performs, in terms of alternating current (AC) and direct current (DC). This leads to four main types of power converters:

AC–DC Conversion from AC to regulated or unregulated DC voltage or current.

DC–DC Conversion from a DC input voltage to a DC output voltage, providing regulation of the output voltage and isolation (optional).

DC–AC Conversion from a DC voltage or current to an AC voltage or current with controlled (variable) amplitude, frequency, and phase.

AC–AC Conversion from an AC voltage with fixed magnitude and frequency to an AC voltage with controlled (variable) amplitude and frequency.

Each one of these types includes several subcategories, as depicted in Figure 1.3. Some examples of different types of power converters will be described and analyzed in this book.

1.3 Control of Power Converters and Drives

Control schemes for power converters and drives have been constantly evolving according to the development of new semiconductor devices and the introduction of new control platforms. While diode rectifiers operate without any control, analog control circuits were introduced for regulating the firing angle of thyristors. With the introduction of power transistors with faster switching frequencies, analog control circuits have been used from the beginning and later were replaced by digital control platforms with the possibility of implementing more advanced control schemes.

1.3.1 Power Converter Control in the Past

In thyristor-based rectifiers the average value of the output voltage can be adjusted by regulating the angle of the firing pulses in relation to the grid voltage. The control circuit for this power converter must detect the zero crossings of the grid voltage and generate the firing pulses according to the desired angle. Figure 1.4 shows the operation of a single-phase thyristor rectifier with a resistive–inductive load. It can be seen that the firing angle α modifies the waveform of the output voltage and, consequently, the average output voltage.

Thyristors switch at fundamental frequency, because their turn-off instant is line dependent and cannot be controlled. However, with the introduction of power transistors like the insulated-gate bipolar transistor (IGBT), hard switching or controlled turn-off is possible, allowing higher switching frequencies. A simple example of a power converter with only one switch is the buck converter. It generates an output voltage whose average value is between zero and the input voltage. This desired voltage is obtained by adjusting the duty cycle of the switch. A simple control for this converter consists of comparing the reference voltage to a triangular waveform. If the reference is higher that the triangular signal, then the switch is turned on, otherwise the switch is turned off. The power circuit and important waveforms are shown in Figure 1.5.

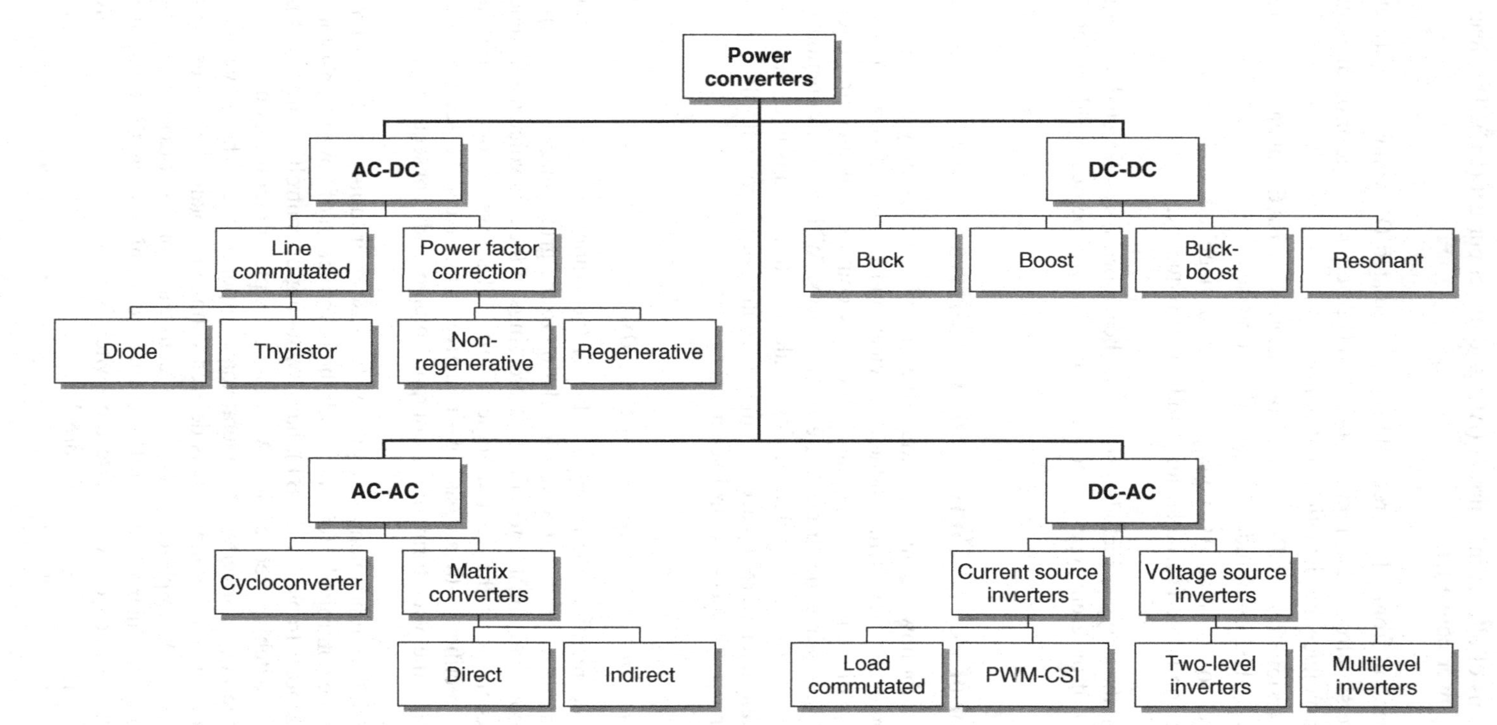

Figure 1.3 Power converter classification (PWM-CSI = Pulse Width-Modulated Current Source Inverter)

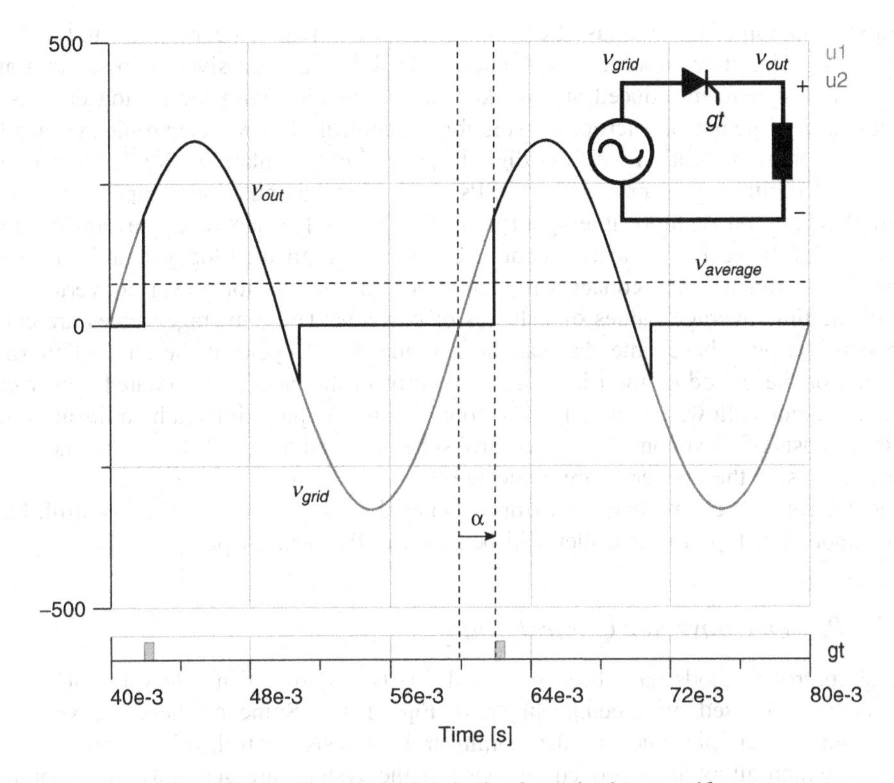

Figure 1.4 Operation of a single-phase thyristor rectifier

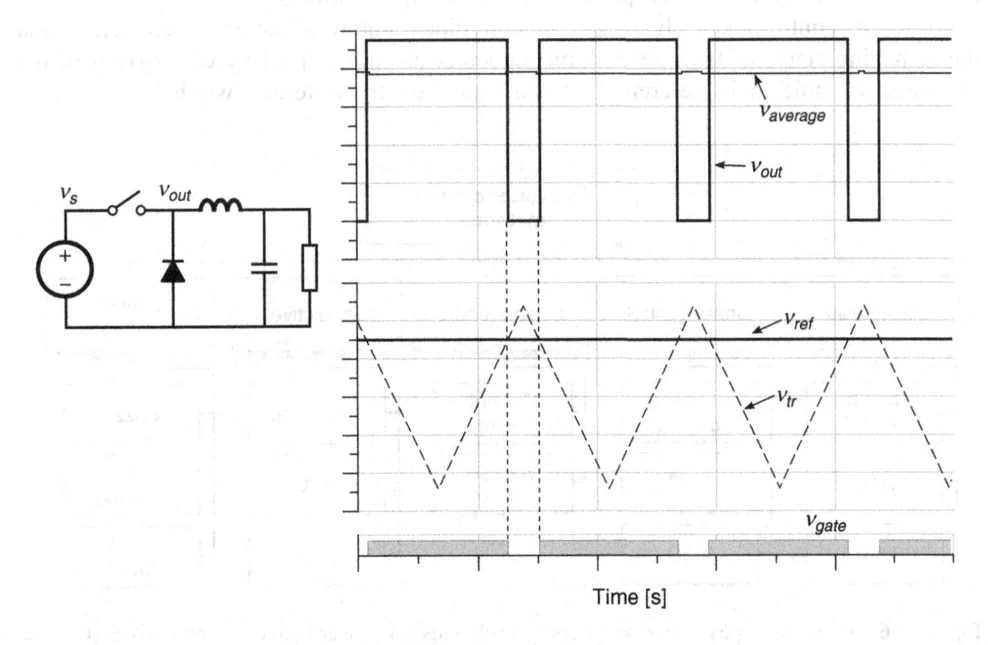

Figure 1.5 Operation of a buck converter

Implementation of these kinds of control schemes was realized completely in the 1960s using analog circuits composed of operational amplifiers and passive components. Later, digital circuits were introduced and worked in combination with the analog circuits. In recent decades, the use of microprocessors for the control of power electronic systems has become a common solution for fully digital control implementations. Modern microcontrollers and digital signal processors (DSPs) with high computational capabilities allow the implementation of more intelligent control schemes [7]. However, several concepts that were developed for the analog control circuits are replicated today in a digital way.

One of the fundamental concepts in analog control circuits for power converters is to control the time-average values of voltages or currents. These average values are calculated considering a base time that can be a fundamental cycle in the case of thyristor rectifiers, or the period of the triangular waveform in the case of modulated power converters. This idea allows the model of the converter to be approximated by a linear system and is the basis of most conventional control schemes used today. However, the nonlinear characteristics of the converter are neglected.

Another control scheme that has its origin in analog circuits is hysteresis control. More details about this type of controller will be given in the next chapter.

1.3.2 Power Converter Control Today

Several control methods have been proposed for the control of inverters and drives, the most commonly used ones being shown in Figure 1.6. Some of these are very well established and simple, such as the nonlinear hysteresis control, while newer control methods, which allow an improved behavior of the system, are generally more complex or need much more calculation power from the control platform.

Hysteresis control takes advantage of the nonlinear nature of the power converters and the switching states of the power semiconductors are determined by comparison of the measured variable to its reference, considering a given hysteresis width for the error.

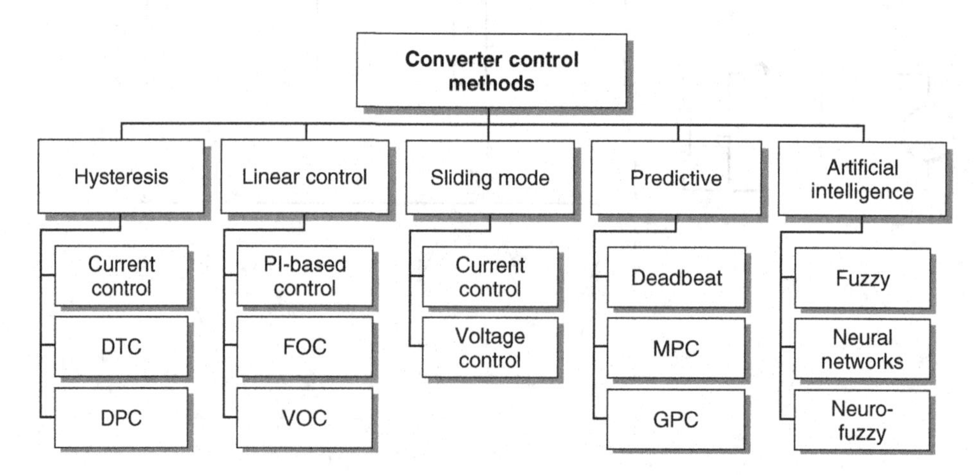

Figure 1.6 Different types of converter control schemes for power converters and drives (GPC = Generalized Predictive Control)

This control scheme can be used in simple applications such as current control, but also for more complex schemes such as direct torque control (DTC) [8] and direct power control (DPC) [9]. This control scheme has its origin in analog electronics, and in order to implement this control scheme in a digital platform, a very high sampling frequency is required. The hysteresis width and the nonlinearity of the system inherently introduce variable switching frequency, which can lead to resonance problems in some applications and generate a spread spectral content. This leads to the need for bulky and expensive filters. Some modifications have been proposed to control the switching frequency.

Given a modulation stage for the converter, any linear controller can be used with the power converters, the most common choice being the use of proportional–integral (PI) controllers. A well-known control method for drives, based on linear controllers, is field-oriented control (FOC) [10, 8]. Similar concepts can be also applied for grid-connected converters with voltage-oriented control (VOC) for the current [11]. The linear control scheme with a modulation stage often requires additional coordinate transformations. In addition, the fact that a linear control is applied to a nonlinear system can lead to uneven performance throughout the dynamic range. Moreover, today's digital implementation requires sampled data control schemes that are an approximation of the continuous-time linear controller. All this, together with the additional modulation stage, introduces several design steps and considerations for achieving a suitable control scheme, which can be very challenging for some power converters such as matrix, multilevel converters, etc. Furthermore, power converter systems are subject to several system constraints and technical requirements (total harmonic distortion (THD), maximum current, maximum switching frequency, etc.), which cannot be directly incorporated into linear controller design. In summary, classical control theory has been adapted over and over in order to use it in modern digitally controlled converters.

With the development of more powerful microprocessors, new control schemes have been proposed. Some of the most important ones are fuzzy logic control, neural networks, sliding mode control, and predictive control.

Among these new control schemes, predictive control appears to be a very interesting alternative for the control of power converters and drives. Predictive control comprises a very wide family of controllers with very different approaches. The common ideas behind all predictive control are the use of a model of the system for calculating predictions of the future behavior of the controlled variables, and the use of an optimization criterion for selecting the proper actuation.

One of the best known predictive control schemes is deadbeat control, which uses a model of the system to calculate the voltage that makes the error zero in one sample time. Then the voltage is applied using a modulator. A different and very powerful predictive control strategy that has been applied quite recently to power electronics is model predictive control (MPC), which is the subject of this book.

1.3.3 Control Requirements and Challenges

Traditionally, control requirements were mainly associated with the dynamic performance and stability of the system. Currently, industry requires more demanding technical specifications and constraints, and in many cases it is subject to regulations and codes. Many of these requirements enforce operating limits and conditions that cannot be dealt with by the hardware only, but also need to be addressed by the control system. This shift in trend has driven the development of more advanced control methods.

The design of an industrial power electronic system can be seen as an optimization problem where several objectives must be fulfilled at the same time. Among these requirements, constraints, and control challenges, the following are especially important in power electronics:

- Provide the smallest possible error in the controlled variables, with fast dynamics for reference following and disturbance rejection.
- Operate the power switches in such a way that switching losses are minimized. This requirement leads to increased efficiency and better utilization of the semiconductor devices.
- Power converters are switched systems that inherently generate harmonic content. Usually this harmonic content is measured as THD. Many power converter systems have limitations and restrictions on the harmonic content introduced by the modulation stage. These limits are usually specified in standards that can change from one country to another.
- The electromagnetic compatibility (EMC) of the system must be considered, according to defined standards and regulations.
- In many systems, common-mode voltages must be minimized due to the harmful effects that they can produce. These voltages induce leakage currents that reduce the safety and lifetime of some systems.
- Good performance for a wide range of operating conditions. Due to the nonlinear nature of power converters, this is difficult to achieve when the controller has been adjusted for a single operating point of the linearized system model.
- Some converter topologies have their own inherent restrictions and constraints such as forbidden switching states, voltage balance issues, power unbalances, mitigation of resonances, and many other specific requirements.

1.3.4 Digital Control Platforms

Control strategies for power converters and drives have been the subject of ongoing research for several decades in power electronics. Classical linear controllers combined with modulation schemes and nonlinear controllers based on hysteresis bounds have become the most used schemes in industrial applications. Many of these concepts go back to research on analog hardware, which limited complexity. Modern digital control platforms like DSPs have become state of the art and have been widely accepted as industrial standards. The main digital control platforms used in industrial electronics are based on fixed-point processor, due to the high computational power and low cost. However, in the academic world, control platforms based on floating-point processor with high programming flexibility are more usually used. Recently, hardware and software solutions implemented in field programmable gate arrays (FPGAs) have received particular attention, mainly because of their ability to allow designers to build efficient and dedicated hardware architectures by means of flexible software. The main stream control platforms used in power electronics are summarized in Table 1.1. An example of the continuously increasing computational power of digital hardware is shown Figure 1.7.

Table 1.1 Examples of digital control platforms

DSP TMS320F2812	DSP TMS320C6713	dSPACE DS1104–DS1103	FPGA XC3S400
150 MHZ Fixed-point 150 MIPS	225 MHz Floating-point 1800 MIPS	350 MHz–1 GHz Floating-point 662–2500 MIPS	50 MHz Fixed-point –

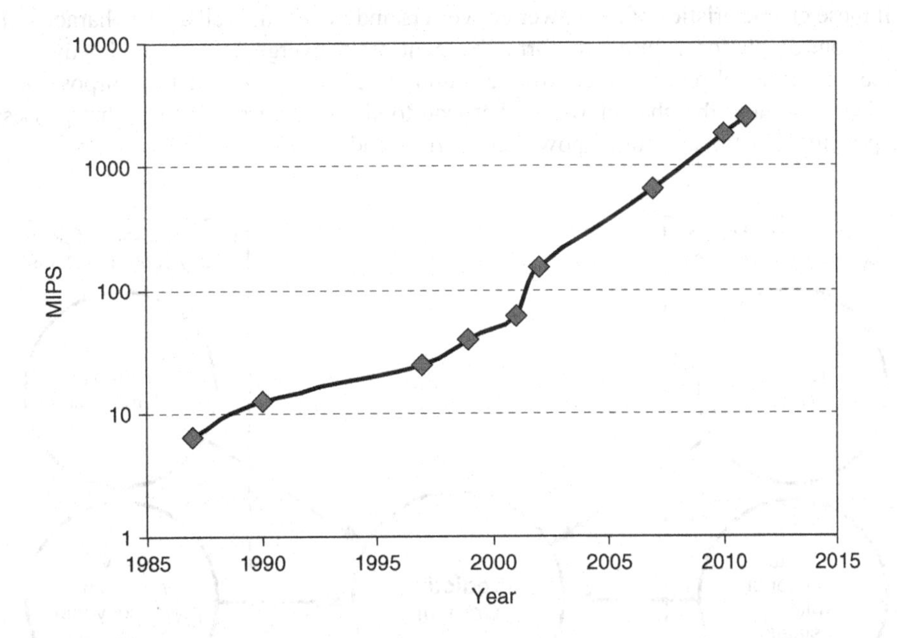

Figure 1.7 Evolution of the processing capabilities of digital hardware

This computational power is measured in terms of millions of instructions per second (MIPS). The high computational power of today's control platforms allows the implementation of new and generally more complex control techniques, for example, fuzzy, adaptive, sliding mode, and predictive control techniques.

1.4 Why Predictive Control is Particularly Suited for Power Electronics

Considering the increasing demands in performance and efficiency of power converters and drives, the development of new control schemes must take into account the real nature of these kinds of systems. Power converters and drives are nonlinear systems of a hybrid nature, including linear and nonlinear parts and a finite number of switching devices. The input signals for power electronic devices are discrete signals that command the turn-on and turn-off transitions of each device. Several constraints and restrictions need to be considered by the control, some of which are inherent to the system, like the maximum output voltage of the inverter, while others are imposed for security reasons, like current limitations to protect the converter and its loads.

Nowadays, practically all control strategies are implemented in digital control platforms running at discrete time steps. Design of any control system must take into account the model of the plant for ajusting the controller parameters, which in the case of power converters and drives is well known. As described in the previous section, control platforms offer an increasing computational capability and more calculation-demanding control algorithms are feasible today. This is the case for predictive control.

All these characteristics of the power converters and drives, as well as the characteristics of the control platforms used to form the control, converge in a natural way to the application of model predictive control, as summarized in Figure 1.8. The purpose of this book is to highlight the characteristics that lead to simple control schemes that possess a high potential for the control of power converters and drives.

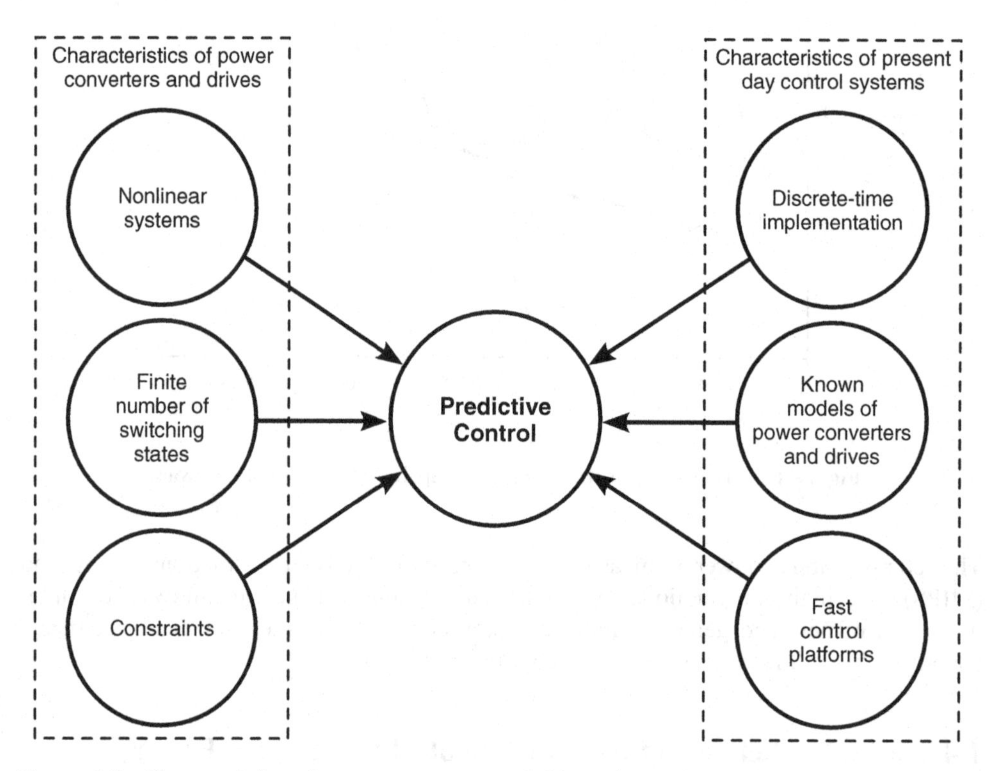

Figure 1.8 Characteristics of power converters and drives that make predictive control a natural solution

1.5 Contents of this Book

The book is organized in four parts. Basic information about power converters, drives, and the classical control schemes appears in the first part. A brief introduction to the basic theory behind model predictive control is also included in this first part. The second part includes several examples of the application of predictive control to different power converter topologies. The third part focuses on predictive control schemes for motor drives, considering induction machines and permanent magnet synchronous motors. The fourth part summarizes several design and implementation aspects that are not covered in the previous parts.

The contents of the subsequent chapters can be summarized as follows:

- Chapter 2 presents some of the most established control methods for current control and state-of-the-art control schemes for drives. These control schemes will be considered as a reference point for comparison of the predictive control schemes presented throughout the book.
- Chapter 3 contains the basic principles of model predictive control and the main considerations that are taken into account for its application to power converters and drives.
- Chapter 4 introduces the application of predictive control to power converters, considering one of the most common converters, the three-phase inverter.
- A three-level neutral-point clamped inverter is considered in Chapter 5. A predictive current control scheme that takes into account the special requirements imposed by this converter topology is presented.
- Different predictive control schemes for active front-end rectifiers are presented in Chapter 6, including current control and power control.
- The application of predictive control to a matrix converter is covered in Chapter 7. Control schemes for the input and output variables are presented.
- Predictive control schemes for induction machines are presented in Chapter 8. Current control and torque control schemes are considered.
- Chapter 9 presents current control and speed control for permanent magnet synchronous motors.
- Considerations on the formulation of an appropriate cost function are discussed in Chapter 10.
- Some guidelines on how to adjust the weighting factors of the cost function are presented in Chapter 11.
- The high number of calculations required by predictive control introduces time delays between the measurements and the actuation. A method to compensate this delay, and related topics, are included in Chapter 12.
- An empirical approach to assess the effect of model parameter errors in the performance of predictive control is presented in Chapter 13.

In the appendices to this book, information about how to implement MATLAB® simulations of predictive control schemes is given, considering three different examples. These examples will allow the reader to start implementing predictive control from proven simulations. In this way, it will be easier for the reader to understand the principles of predictive control and then adapt these schemes to different applications.

References

[1] N. Mohan, T. M. Undeland, and W. P. Robbins, *Power electronics*, 3rd ed. John Wiley & Sons, Inc., 2003.

[2] B. Bose, "Energy, environment, and advances in power electronics," *IEEE Transactions on Power Electronics*, vol. 15, no. 4, pp. 688–701, July 2000.

[3] J. Rodríguez, J. Pontt, N. Becker, and A. Weinstein, "Regenerative drives in the megawatt range for high-performance downhill belt conveyors," *IEEE Transactions on Industry Applications*, vol. 38, no. 1, pp. 203–210, January/February 2002.

[4] J. Rodríguez, J. Pontt, G. Alzamora *et al.*, "Novel 20-MW downhill conveyor system using three-level converters," *IEEE Transactions on Industry Applications*, vol. 49, no. 5, pp. 1093–1100, October 2002.

[5] "Grid code: High and extra high voltage," E.ON Netz GmbH, Bayreuth, Germany, Status: 1 April 2006.

[6] D. Ionel, "High-efficiency variable-speed electric motor drive technologies for energy savings in the US residential sector," in 12th International Conference on Optimization of Electrical and Electronic Equipment (OPTIM), pp. 1403–1414, May 2010.

[7] G.-A. Capolino, "Recent advances and applications of power electronics and motor drives–advanced and intelligent control techniques," in 34th Annual IEEE Conference on Industrial Electronics, IECON. pp. 37–39, November 2008.

[8] I. Takahashi and T. Noguchi, "A new quick response and high efficiency control strategy for an induction motor," *IEEE Transactions on Industry Applications*, vol. 22, no. 5, pp. 820–827, September 1986.

[9] T. Ohnishi, "Three phase PWM converter/inverter by means of instantaneous active and reactive power control," in Proceedings of the International Conference on Industrial Electronics, Control and Instrumentation, 1991. Proceedings. IECON '91. vol. 1, pp. 819–824, October–November 1991.

[10] F. Blaschke, "The principle of field-orientation applied to transvector closed-loop control system for rotating field machines," *Siemens Review*, vol. XXXIX, no. 5, pp. 217–219, 1972.

[11] M. Kazmierkowski and L. Malesani, "Current control techniques for three-phase voltage-source pwm converters: a survey," *IEEE Transactions on Industrial Electronics*, vol. 45, no. 5, pp. 691–703, October 1998.

2

Classical Control Methods for Power Converters and Drives

The use of power converters has become very popular in the last few decades for a wide range of applications, including drives, energy conversion, traction, and distributed generation. The control of power converters has been extensively studied and new control schemes are presented every year.

Several control schemes have been proposed for the control of power converters and drives. Some of them are shown in Figure 1.6 in the previous chapter. From these, hysteresis control and linear control with pulse width modulation are the most established in the literature [1–3]. However, with the development of faster and more powerful microprocessors, implementation of new and more complex control schemes is possible. Some of these new control schemes for power converters include fuzzy logic, sliding mode control, and predictive control. Fuzzy logic is suitable for applications where the controlled system or some of its parameters are unknown. Sliding mode control presents robustness and takes into account the switching nature of the power converters. Other control schemes found in the literature include neural networks, neuro-fuzzy, and other advanced control techniques.

Predictive control presents several advantages that make it suitable for the control of power converters: the concepts are intuitive and easy to understand; it can be applied to a variety of systems; constraints and nonlinearities can be easily included; multivariable cases can be considered; and the resulting controller is easy to implement. It requires a high number of calculations, compared to a classical control scheme, but the fast microprocessors available today make possible the implementation of predictive control. Generally, the quality of the controller depends on the quality of the model.

2.1 Classical Current Control Methods

One of the most studied topics in power converters control is current control, for which there are two classical control methods that have been extensively studied over the last few decades: namely, hysteresis control and linear control using pulse width modulation (PWM) [4, 1, 2].

Predictive Control of Power Converters and Electrical Drives, First Edition. Jose Rodriguez and Patricio Cortes.

2.1.1 Hysteresis Current Control

The basic idea of hysteresis current control is to keep the current inside the hysteresis band by changing the switching state of the converter each time the current reaches the boundary. Figure 2.1 shows the hysteresis control scheme for a single-phase inverter. Here, the current error is used as the input of the comparator and if the current error is higher than the upper limit $\delta/2$, the power switches T_1, T_4 are turned on and T_2, T_3 are turned off. The opposite switching states are generated if the error is lower than $-\delta/2$. It can be observed in Figure 2.1 that with this very simple strategy the load current i_L follows the waveform of the reference current i_L^* very well, which in this case is sinusoidal.

For a three-phase inverter, measured load currents of each phase are compared to the corresponding references using hysteresis comparators, as shown in Figure 2.2. Each comparator determines the switching state of the respective inverter leg (S_a, S_b, and S_c) such that the load currents are forced to remain within the hysteresis band. Due to the interaction between the phases, the current error is not strictly limited to the value of the hysteresis band. A simplified diagram of this control strategy is shown in Figure 2.3.

This method is conceptually simple and the implementation does not require complex circuits or processors. The performance of the hysteresis controller is good, with a fast dynamic response. The switching frequency changes according to variations in the hysteresis width, load parameters, and operating conditions. This is one of the major drawbacks of hysteresis control, since variable switching frequency can cause resonance

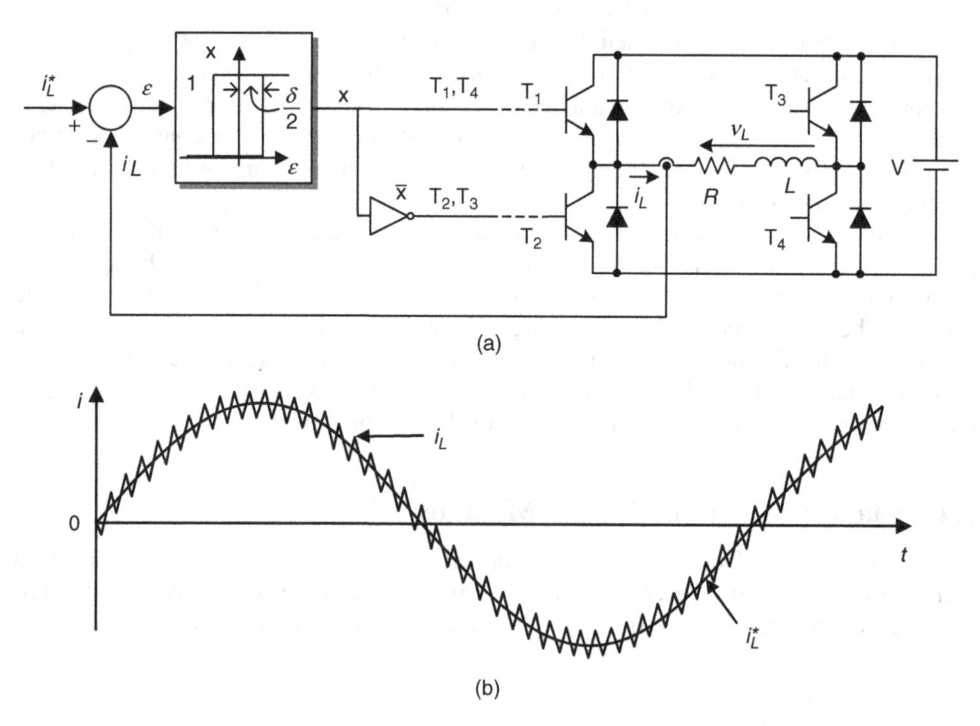

Figure 2.1 Hysteresis current control for a single-phase inverter. (a) Control scheme. (b) Load current

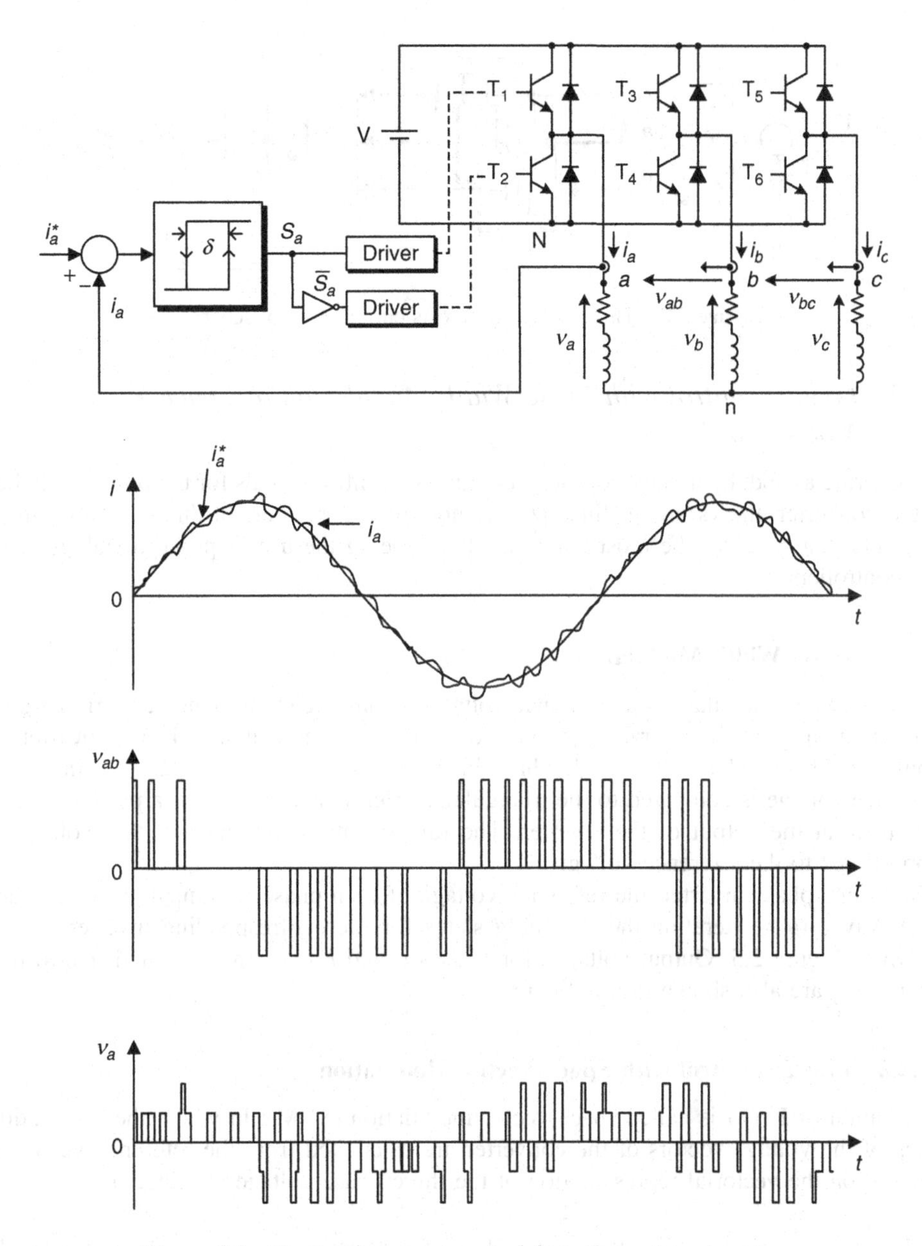

Figure 2.2 Hysteresis current control for a three-phase inverter

problems. In addition, the switching losses restrict the application of hysteresis control to lower power levels [4]. Several modifications have been proposed in order to control the switching frequency of the hysteresis controller.

When implemented in a digital control platform, a very high sampling frequency is required in order to keep the controlled variables within the hysteresis band all the time.

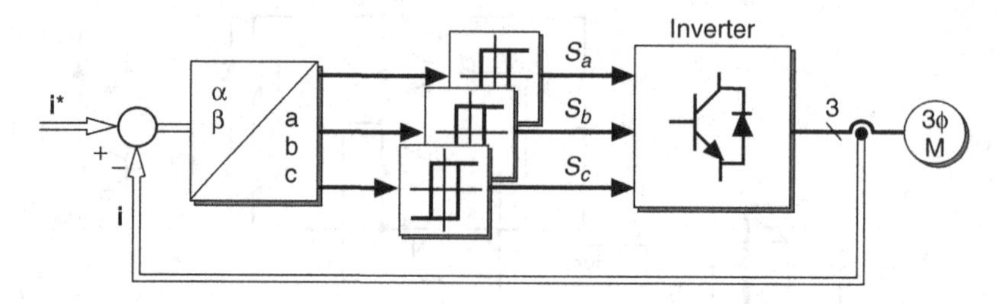

Figure 2.3 Three-phase hysteresis current control scheme

2.1.2 Linear Control with Pulse Width Modulation or Space Vector Modulation

Considering a modulator stage for the generation of control signals for the power switches of the converter allows one to linearize the nonlinear converter. In this way, any linear controller can be used, the most common choice being the use of proportional–integral (PI) controllers.

2.1.2.1 Pulse Width Modulation

In a pulse width modulator, the reference voltage is compared to a triangular carrier signal and the output of the comparator is used to drive the inverter switches. The application of a pulse width modulator in a single-phase inverter is shown in Figure 2.4. A sinusoidal reference voltage is compared to the triangular carrier signal generating a pulsed voltage waveform at the output of the inverter. The fundamental component of this voltage is proportional to the reference voltage.

In a three-phase inverter, the reference voltage of each phase is compared to the triangular waveform, generating the switching states for each corresponding inverter leg, as shown in Figure 2.5. Output voltages for phases a and b, v_{aN} and v_{bN}, and line-to-line voltage v_{ab} are also shown in this figure.

2.1.2.2 Linear Control with Space Vector Modulation

A variation of PWM is called space vector modulation (SVM), in which the application times of the voltage vectors of the converter are calculated from the reference vector. It is based on the vectorial representation of the three-phase voltages, defined as

$$\mathbf{v} = \frac{2}{3}(v_{aN} + \mathbf{a}v_{bN} + \mathbf{a}^2 v_{cN}) \tag{2.1}$$

where v_{aN}, v_{bN}, and v_{cN} are the phase-to-neutral (N) voltages of the inverter and $\mathbf{a} = e^{j2[\pi]/3}$. The output voltages of the inverter depend on the switching state of each phase and the DC link voltage, $v_{xN} = S_x V_{dc}$, with $x = \{a, b, c\}$. Then, taking into account the combinations of the switching states of each phase, the three-phase inverter generates the voltage vectors listed in Table 2.1 and depicted in Figure 2.6.

Considering the voltage vectors generated by the inverter, the $\alpha - \beta$ plane is divided into six sectors, as shown in Figure 2.6. In this way, a given reference voltage vector

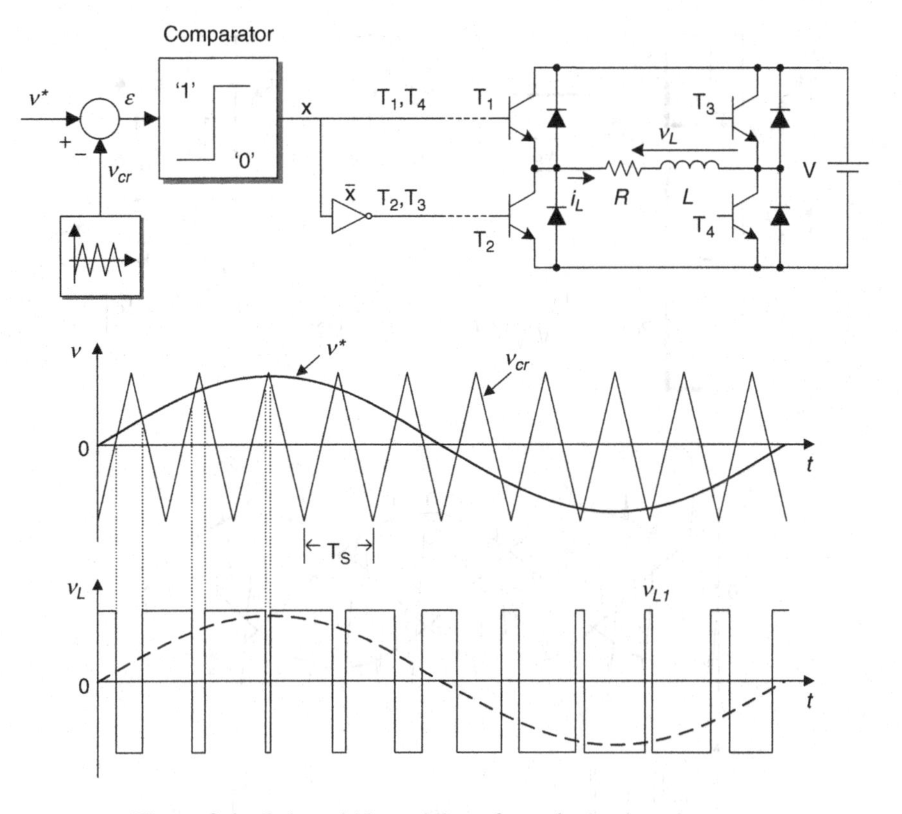

Figure 2.4 Pulse width modulator for a single-phase inverter

$\mathbf{v}^*$, located at a generic sector k, can be synthesized using the adjacent vectors $\mathbf{V}_k$, $\mathbf{V}_{k+1}$, and $\mathbf{V}_0$, applied during t_k, t_{k+1}, and t_0, respectively. This can be summarized with the following equations:

$$\mathbf{v}^* = \frac{1}{T}(\mathbf{V}_k t_k + \mathbf{V}_{k+1} t_{k+1} + \mathbf{V}_0 t_0) \tag{2.2}$$

$$T = t_k + t_{k+1} + t_0 \tag{2.3}$$

where T is the carrier period and t_k/T, t_{k+1}/T, and t_0/T are the duty cycles of their respective vectors. Using trigonometric relations the application time for each vector can be calculated, resulting in

$$t_k = \frac{3T|\mathbf{v}^*|}{2V_{dc}}\left(\cos(\theta - \theta_k) - \frac{\sin(\theta - \theta_k)}{\sqrt{3}}\right) \tag{2.4}$$

$$t_{k+1} = \frac{3T|\mathbf{v}^*|}{V_{dc}}\frac{\sin(\theta - \theta_k)}{\sqrt{3}} \tag{2.5}$$

$$t_0 = T - t_k - t_{k+1} \tag{2.6}$$

where θ is the angle of the reference vector $\mathbf{v}^*$ and θ_k is the angle of vector $\mathbf{V}_k$.

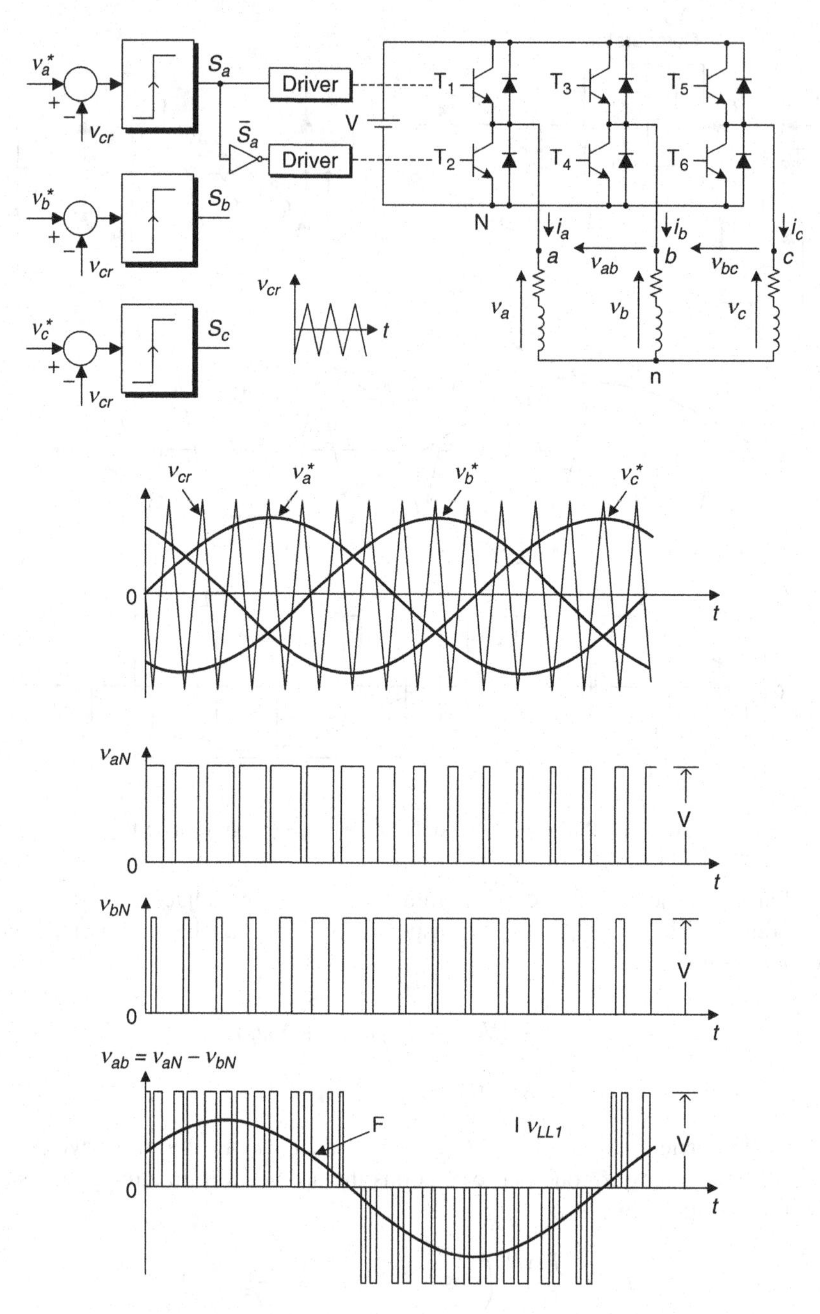

Figure 2.5 Pulse width modulator for a three-phase inverter

Table 2.1 Switching states and voltage vectors

S_a	S_b	S_c	Voltage vector **V**
0	0	0	$\mathbf{V}_0 = 0$
1	0	0	$\mathbf{V}_1 = \frac{2}{3}V_{dc}$
1	1	0	$\mathbf{V}_2 = \frac{1}{3}V_{dc} + j\frac{\sqrt{3}}{3}V_{dc}$
0	1	0	$\mathbf{V}_3 = -\frac{1}{3}V_{dc} + j\frac{\sqrt{3}}{3}V_{dc}$
0	1	1	$\mathbf{V}_4 = -\frac{2}{3}V_{dc}$
0	0	1	$\mathbf{V}_5 = -\frac{1}{3}V_{dc} - j\frac{\sqrt{3}}{3}V_{dc}$
1	0	1	$\mathbf{V}_6 = \frac{1}{3}V_{dc} - j\frac{\sqrt{3}}{3}V_{dc}$
1	1	1	$\mathbf{V}_7 = 0$

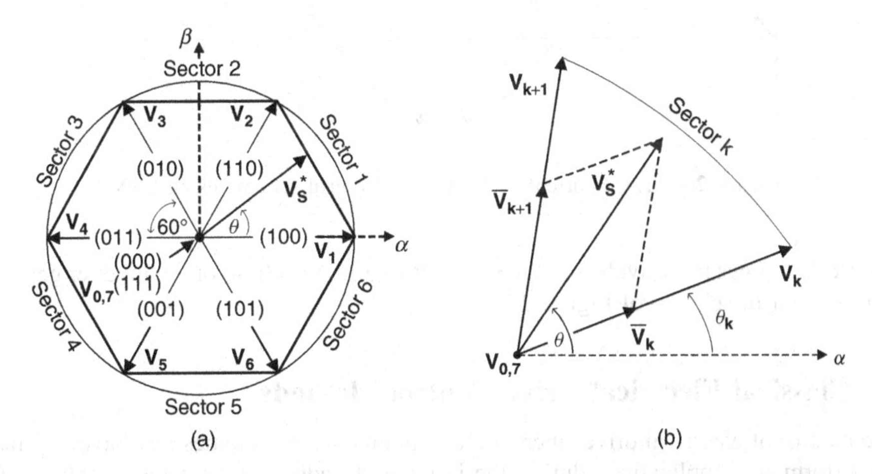

Figure 2.6 Principles of space vector modulation (SVM). (a) Voltage vectors and sector definition. (b) Generation of the reference vector in a generic sector

A classical current control scheme using SVM is shown in Figure 2.7. Here, the error between the reference and the measured load current is processed by a PI controller to generate the reference load voltages.

With this method, constant switching frequency, fixed by the carrier, is obtained. The performance of this control scheme depends on the design of the controller parameters and on the frequency of the reference current. Although the PI controller assures zero steady state error for continuous reference, it can present a noticeable error for sinusoidal references. This error increases with the frequency of the reference current and may become unacceptable for certain applications [4]. To overcome the problem of the PI controller with sinusoidal references, the standard solution is to modify the original scheme considering a coordinate transformation to a rotating reference frame in which the reference currents are constant values.

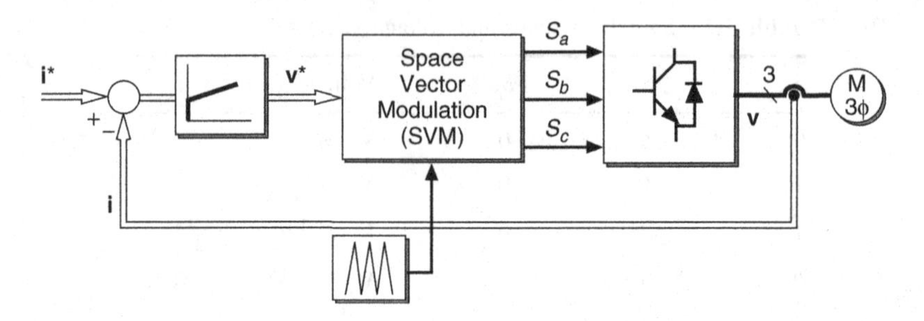

Figure 2.7 Classical control scheme using SVM

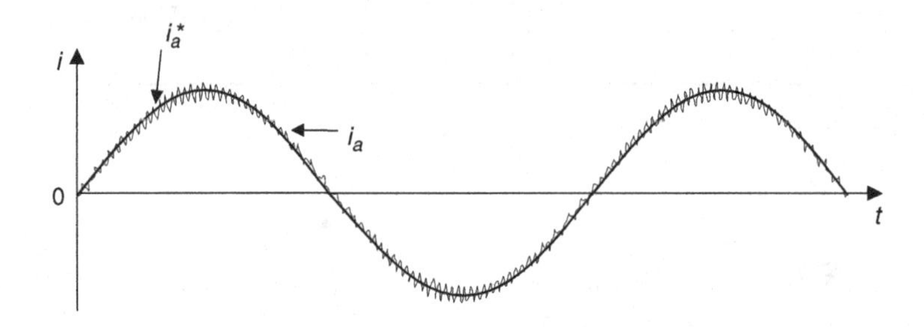

Figure 2.8 Load current for a classical control scheme using SVM

Figure 2.8 shows the waveform of the load current in one phase of the inverter generated using the control scheme of Figure 2.7.

2.2 Classical Electrical Drive Control Methods

In the control of electrical drives there are two main control schemes that have dominated high-performance applications during the last few decades: field-oriented control (FOC) [5–7] and direct torque control (DTC) [8]. These control strategies will be presented in the next two sections and will be considered for comparison to the predictive control schemes for electrical drives presented later in this book.

2.2.1 Field Oriented Control

The main idea behind FOC is the use of a proper coordinate system that allows decoupled control over the electrical torque T_e and the magnitude of the rotor flux $|\boldsymbol{\Psi}_r|$. This can be achieved by aligning the coordinate system with the rotor flux.

Figure 2.9 shows the relation between the stationary $\alpha\beta$ and rotating reference frame dq, which is aligned with the rotor flux vector $\boldsymbol{\Psi}_r$.

Since the variables are expressed in a rotating coordinates frame, the electromagnetic torque can be controlled via the imaginary component of the stator current i_{sq} and the rotor flux magnitude is controlled by its real part i_{sd}. These relations are obtained from

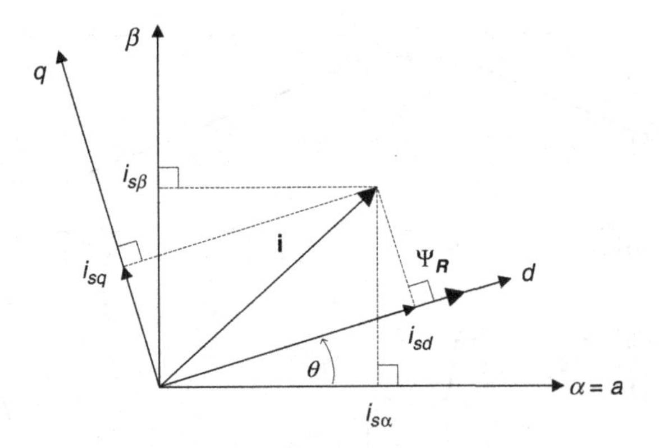

Figure 2.9 Reference frame orientation in FOC

the machine model expressed in the rotating coordinates frame:

$$\psi_{rd} = \frac{L_m}{\tau_r s + 1} i_{sd} \tag{2.7}$$

$$T_e = \frac{3}{2} \frac{L_m}{L_r} p \psi_{rd} i_{sq} \tag{2.8}$$

where τ_r is the time constant of the rotor [7].

A block diagram for FOC is shown in Figure 2.10 where the reference current i_{sq}^* is obtained from an outer speed control loop while i_{sd}^* is obtained from the rotor flux control loop. The stator current errors are controlled using PI controllers which generate the stator reference voltages v_{sd}^* and v_{sq}^*. Then, these voltages are converted to the stationary reference frame and applied to the inverter using a pulse width modulator.

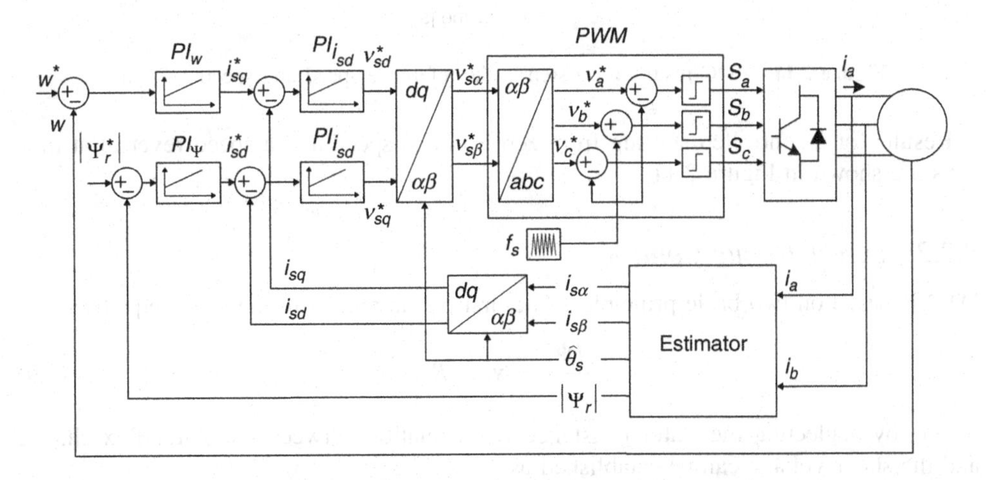

Figure 2.10 FOC block diagram

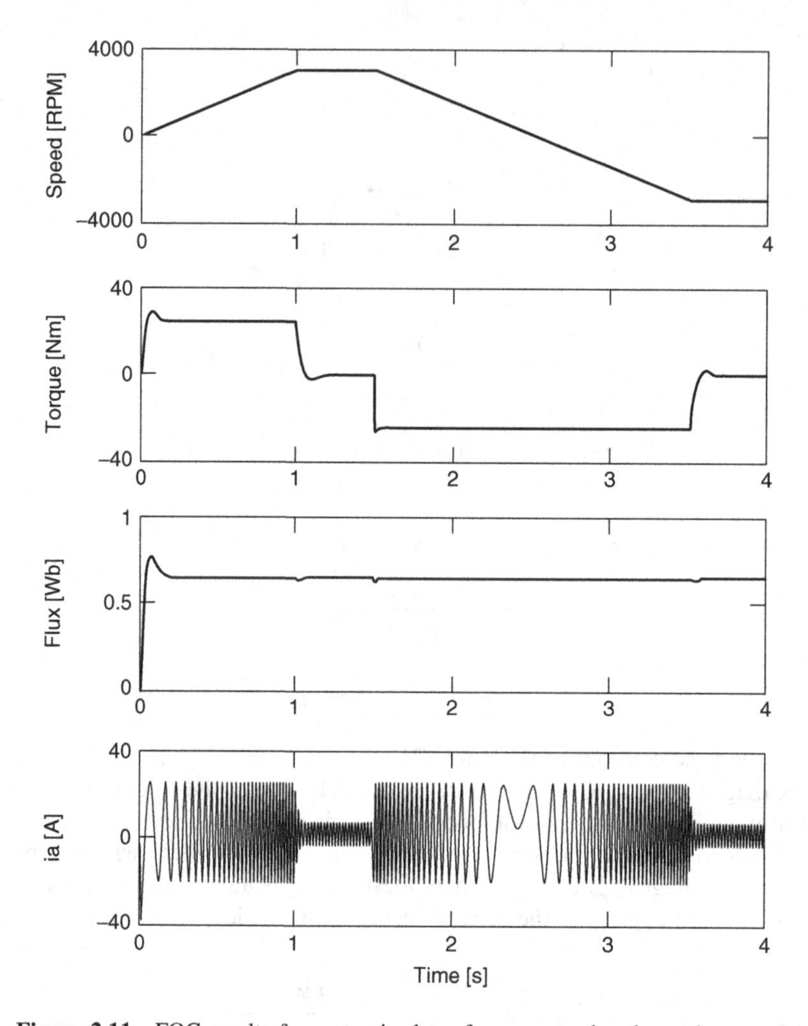

Figure 2.11 FOC results for a step in the reference speed and speed reversal

Results for a controlled starting from zero to rated speed and a speed reversal at time 1.5 s are shown in Figure 2.11.

2.2.2 *Direct Torque Control*

DTC is based on two basic principles. The first one is related to the stator equation

$$\frac{d\boldsymbol{\Psi}_s}{dt} = \mathbf{v}_s - R_s\mathbf{i}_s \tag{2.9}$$

where, by neglecting the stator resistance R_s, a relation between the stator flux change and the stator voltage can be established as

$$\Delta\boldsymbol{\Psi}_s = \boldsymbol{\Psi}_s(t + T_s) - \boldsymbol{\Psi}(t) \approx \mathbf{v}_s T_s \tag{2.10}$$

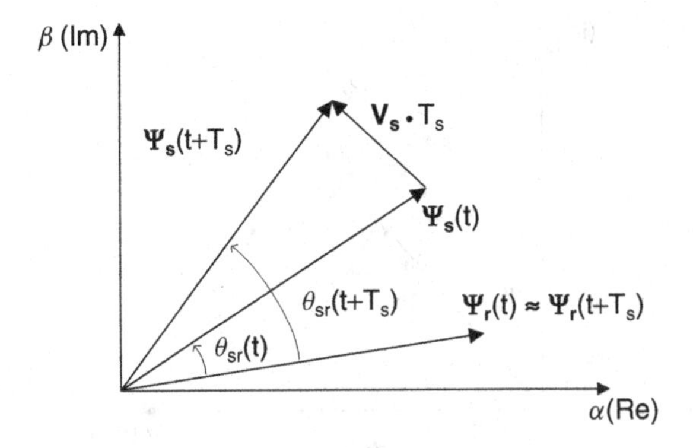

Figure 2.12 Principles of DTC. Stator and rotor flux vectors

Hence, the stator flux can be changed by the application of a given stator voltage vector during a time T_s. This allows control of the stator flux vector, making it follow a given trajectory.

The second assumption is that the rotor flux dynamics are slower than the dynamics of the stator flux. It can be assumed that during one sample time interval the rotor flux vector remains invariant. Besides, it has been demonstrated that the electromagnetic torque T_e depends on the angle θ_{sr} between the stator and rotor flux vectors [8]:

$$T_e = \frac{3}{2} p \frac{L_m}{L_s L_r - L_m^2} |\boldsymbol{\Psi}_s| |\boldsymbol{\Psi}_r| \sin(\theta_{sr}) \tag{2.11}$$

As shown in Figure 2.12, the angle θ_{sr} can be changed by the application of the proper stator voltage vector $\mathbf{v}_s$.

Considering the voltage vectors generated by a two-level inverter, the complex plane is divided into six sectors, as shown in Figure 2.13. Then for each sector the effect of each voltage vector on the behavior of the torque and flux is evaluated. For example, if

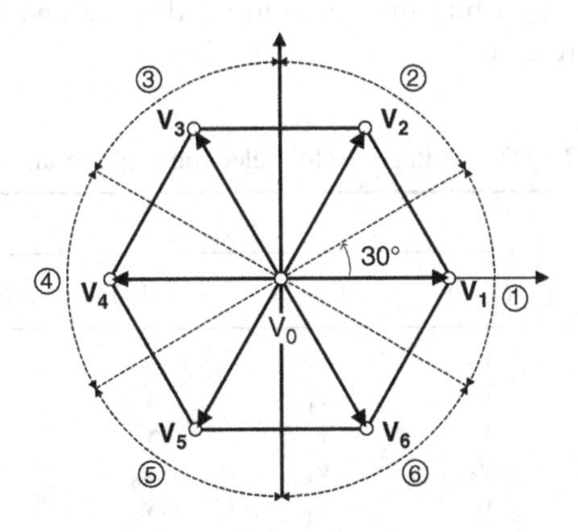

Figure 2.13 Definition of sectors for DTC

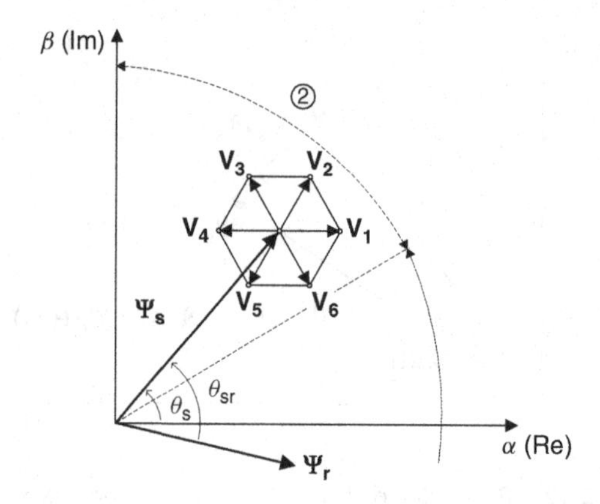

Figure 2.14 Example for voltage vector selection in DTC

the stator flux vector $\boldsymbol{\Psi}_s$ is located in sector 2, as shown in Figure 2.14, applying vector $\mathbf{V}_3$ will increase T_e and $|\boldsymbol{\Psi}_s|$, while applying $\mathbf{V}_1$ will decrease T_e and increase $|\boldsymbol{\Psi}_s|$. In this way, a look-up table is compiled, considering the increase or decrease of T_e and $|\boldsymbol{\Psi}_s|$ for each sector.

The resulting look-up table for DTC is presented in Table 2.2. The inputs of the table are the sector of the stator flux, defined in Figure 2.13, and the control signals h_Ψ and h_T, which define the required increase ("1") or decrease ("−1") of the stator flux magnitude and the electrical torque, respectively.

A block diagram of DTC is depicted in Figure 2.15. An external speed control loop generates the reference for the torque T_e^*, while the reference for the magnitude of the stator flux is constant. The machine model is used for estimating the torque and the magnitude and angle of the stator flux vector. Torque and flux errors are controlled using individual hysteresis comparators. The output of these comparators, h_T and h_Ψ, and the stator flux angle θ_s are the inputs of the voltage vector selection look-up table. The selected voltage vector is directly applied to the inverter and the machine responds to the control action according to the DTC principle.

Results for a controlled starting from zero to rated speed and a speed reversal at time 1.5 s are shown in Figure 2.16.

Table 2.2 DTC voltage vector selection look-up table

	(h_Ψ, h_T)			
Sector	(1, 1)	(1, −1)	(−1, 1)	(−1, −1)
1	$\mathbf{v}_2$	$\mathbf{v}_6$	$\mathbf{v}_3$	$\mathbf{v}_5$
2	$\mathbf{v}_3$	$\mathbf{v}_1$	$\mathbf{v}_4$	$\mathbf{v}_6$
3	$\mathbf{v}_4$	$\mathbf{v}_2$	$\mathbf{v}_5$	$\mathbf{v}_1$
4	$\mathbf{v}_5$	$\mathbf{v}_3$	$\mathbf{v}_6$	$\mathbf{v}_2$
5	$\mathbf{v}_6$	$\mathbf{v}_4$	$\mathbf{v}_1$	$\mathbf{v}_3$
6	$\mathbf{v}_1$	$\mathbf{v}_5$	$\mathbf{v}_2$	$\mathbf{v}_4$

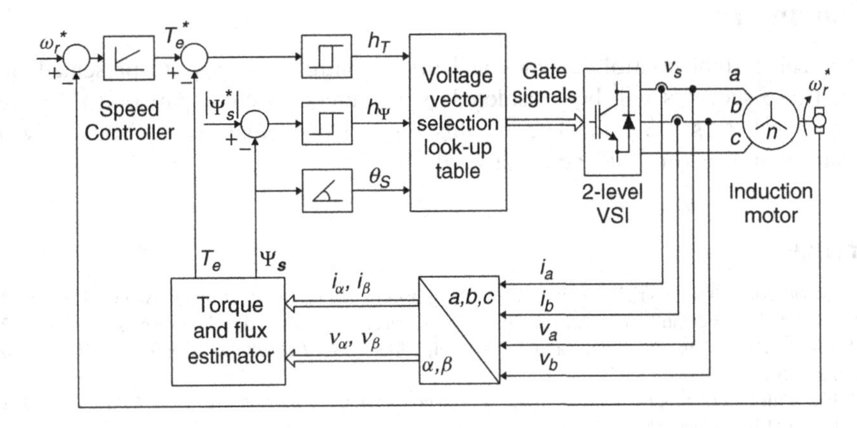

Figure 2.15 Block diagram of DTC (VSI = Voltage Source Inverter)

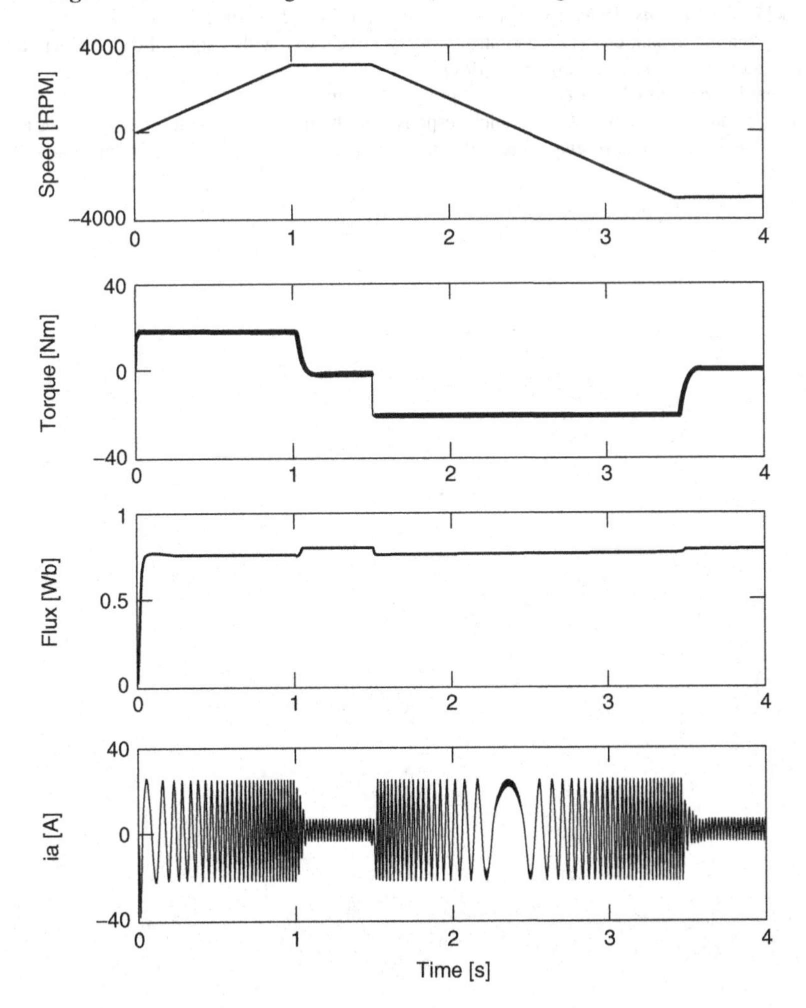

Figure 2.16 DTC results for a step in the reference speed and speed reversal

2.3 Summary

The classical current control schemes and drive control schemes are presented in this chapter. These schemes will be considered as a reference for comparison to the predictive control schemes presented later in this book. Conceptual comparisons as well as performance comparisons will be considered.

References

[1] M. P. Kazmierkowski, R. Krishnan, and F. Blaabjerg, *Control in power electronics*. Academic Press, 2002.
[2] N. Mohan, T. M. Undeland, and W. P. Robbins, *Power electronics*, 3rd ed. John Wiley & Sons, Inc., 2003.
[3] A. Linder, R. Kanchan, P. Stolze, and R. Kennel, *Model-Based Predictive Control of Electric Drives*. Cuvillier Verlag, 2010.
[4] J. Holtz, "Pulsewidth modulation electronic power conversion," *Proceedings of the IEEE*, vol. 82, no. 8, pp. 1194–1214, August 1994.
[5] F. Blaschke, "The principle of field-orientation applied to transvector closed-loop control system for rotating field machines," *Siemens Review*, vol. XXXIX, no. 5, pp. 217–219, 1972.
[6] K. Hasse, "On the dynamics of speed control of a static AC drive with a squirrel-cage induction machine," PhD dissertation, Hochschule Darmstadt, 1969.
[7] W. Leonhard, *Control of Electrical Drives*. Springer Verlag, 1996.
[8] I. Takahashi and T. Noguchi, "A new quick response and high efficiency control strategy for an induction motor," *IEEE Transactions on Industry Applications*, vol. 22, no. 5, pp. 820–827, September 1986.

3

Model Predictive Control

3.1 Predictive Control Methods for Power Converters and Drives

Predictive control covers a very wide class of controllers that have found rather recent application in power converters. A classification for different predictive control methods is shown in Figure 3.1, as proposed in [1].

The main characteristic of predictive control is the use of a model of the system for predicting the future behavior of the controlled variables. This information is used by the controller to obtain the optimal actuation, according to a predefined optimization criterion.

The optimization criterion in hysteresis-based predictive control is to keep the controlled variable within the boundaries of a hysteresis area [2], while in trajectory-based control the variables are forced to follow a predefined trajectory [3]. In deadbeat control, the optimal actuation is the one that makes the error equal to zero in the next sampling instant [4, 5]. A more flexible criterion is used in model predictive control (MPC), expressed as a cost function to be minimized [6].

The difference between these groups of controllers is that deadbeat control and MPC with continuous control set need a modulator in order to generate the required voltage. This will result in having a fixed switching frequency. The other controllers directly generate the switching signals for the converter, do not need a modulator, and will present a variable switching frequency.

One advantage of predictive control is that concepts are very simple and intuitive. Depending on the type of predictive control, implementation can also be simple, as with deadbeat control and finite control set MPC (especially for a two-level converter with horizon $N = 1$). However, some implementations of MPC can be more complex if the continuous control set is considered. Variations of the basic deadbeat control, in order to make it more robust, can also become very complex and difficult to understand.

Using predictive control it is possible to avoid the cascaded structure which is typically used in a linear control scheme, obtaining very fast transient responses. An example of this is speed control using trajectory-based predictive control.

Nonlinearities in the system can be included in the model, avoiding the need to linearize the model for a given operating point, and improving the operation of the system for all

Predictive Control of Power Converters and Electrical Drives, First Edition. Jose Rodriguez and Patricio Cortes.

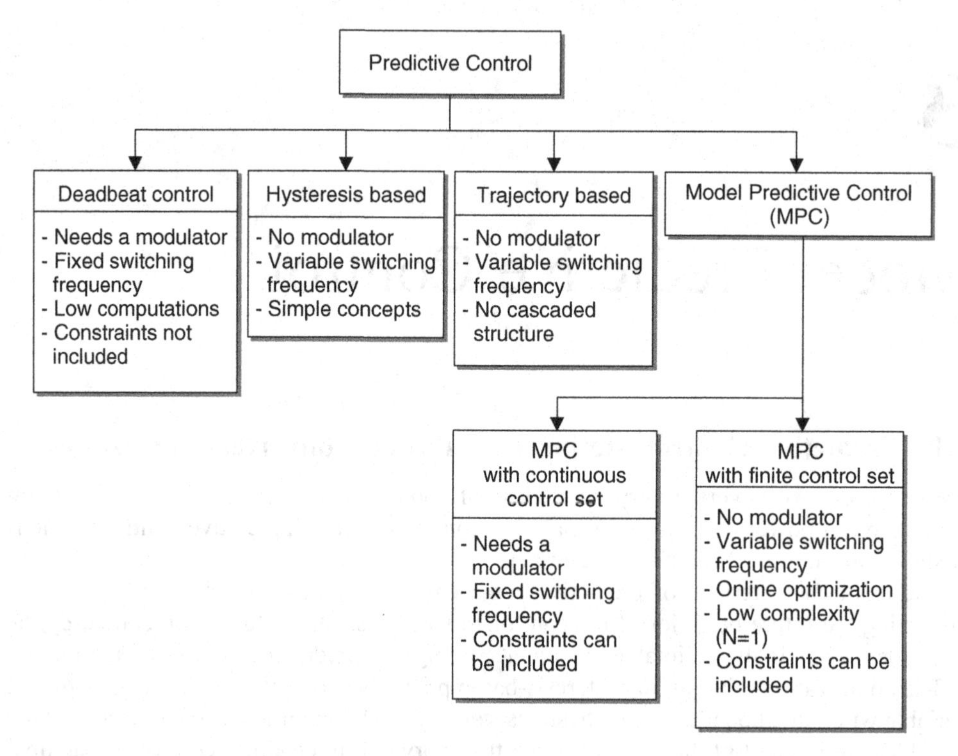

Figure 3.1 Classification of predictive control methods used in power electronics (Cortes *et al.*, 2008 © IEEE)

conditions. It is also possible to include restrictions on some variables when designing the controller. These advantages can be very easily implemented in some control schemes, such as MPC, but are very difficult to obtain in schemes like deadbeat control.

This book will focus on the application of MPC to power converters and drives, considering a finite control set and finite prediction horizon. More details will be found in the following chapters.

3.2 Basic Principles of Model Predictive Control

Among the advanced control techniques, that is, more advanced than standard PID control, MPC is one that has been successfully used in industrial applications [7–9]. Although the ideas of MPC were developed in the 1960s as an application of optimal control theory, industrial interest in these ideas started in the late 1970s [10]. Since then, MPC has been successfully applied in the chemical process industry, where time constants are long enough to perform all the required calculations. Early applications of the ideas of MPC in power electronics can be found from the 1980s considering high-power systems with low switching frequency [2]. The use of higher switching frequencies was not possible at that time due to the large calculation time required for the control algorithm. However, with the development of fast and powerful microprocessors, interest in the application of MPC in power electronics has increased considerably over the last decade.

MPC describes a wide family of controllers, not a specific control strategy [7]. The common elements of this kind of controller are that it uses a model of the system to predict the future behavior of the variables until a predefined horizon in time, and selection of the optimal actuations by minimizing a cost function. This structure has several important advantages:

- Concepts are very intuitive and easy to understand.
- It can be applied to a great variety of systems.
- The multivariable case can be easily considered.
- Dead times can be compensated.
- Easy inclusion of non linearities in the model.
- Simple treatment of constraints.
- The resulting controller is easy to implement.
- This methodology is suitable for the inclusion of modifications and extensions depending on specific applications.

However, some disadvantages have to be mentioned, like the larger number of calculations, compared to classic controllers. The quality of the model has a direct influence on the quality of the resulting controller, and if the parameters of the system change in time, some adaptation or estimation algorithm has to be considered.

The basic ideas present in MPC are:

- The use of a model to predict the future behavior of the variables until a horizon in time.
- A cost function that represents the desired behavior of the system.
- The optimal actuation is obtained by minimizing the cost function.

The model used for prediction is a discrete-time model which can be expressed as a state space model as follows:

$$\mathbf{x}(k+1) = A\mathbf{x}(k) + B\mathbf{u}(k) \tag{3.1}$$

$$\mathbf{y}(k) = C\mathbf{x}(k) + D\mathbf{u}(k) \tag{3.2}$$

A cost function that represents the desired behavior of the system needs to be defined. This function considers the references, future states, and future actuations:

$$J = f(\mathbf{x}(k), \mathbf{u}(k), \ldots, \mathbf{u}(k+N)) \tag{3.3}$$

MPC is an optimization problem that consist of minimizing the cost function J, for a predefined horizon in time N, subject to the model of the system and the restrictions of the system. The result is a sequence of N optimal actuations. The controller will apply only the first element of the sequence

$$\mathbf{u}(k) = [1\ 0\ \cdots\ 0] \arg\min_u J \tag{3.4}$$

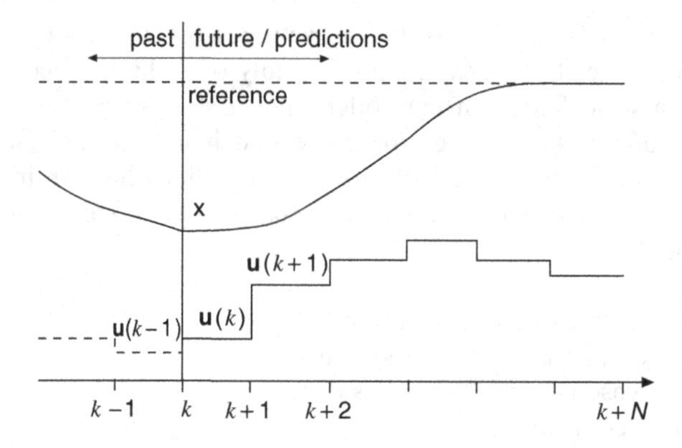

Figure 3.2 Working principle of MPC

where the optimization problem is solved again each sampling instant, using the new measured data and obtaining a new sequence of optimal actuations each time. This is called a *receding horizon* strategy.

The working principle of MPC is summarized in Figure 3.2. The future values of the states of the system are predicted until a predefined horizon in time $k + N$ using the system model and the available information (measurements) until time k. The sequence of optimal actuations is calculated by minimizing the cost function and the first element of this sequence is applied. This whole process is repeated again for each sampling instant considering the new measured data.

3.3 Model Predictive Control for Power Electronics and Drives

Although the theory of MPC was developed in the 1970s, its application in power electronics and drives is more recent due to the fast sampling times that are required in these systems. The fast microcontrollers available in the last decade have triggered research in new control schemes, such as MPC, for power electronics and drives.

As mentioned previously, MPC includes a very wide family of controllers and several different implementations have been proposed. An interesting alternative is the use of generalized predictive control (GPC), which allows solution of the optimization problem analytically, when the system is linear and there are no constraints, providing an explicit control law that can be easily implemented [11, 12]. This control scheme has been used in several power converter [13–15] and drive applications [16–18].

In order to make possible the implementation of MPC in a real system, considering the little time available for calculations due to the fast sampling, it has been proposed to move most of the optimization problem offline using a strategy called explicit MPC. The optimization problem of MPC is solved offline considering the system model, constraints, and objectives, resulting in a look-up table containing the optimal solution as a function of the state of the system. Explicit MPC has been applied for the control of power converters such as DC–DC converters and three-phase inverters [19, 20], and in the control of permanent magnet synchronous motors [21].

Most GPC and explicit MPC schemes approximate the model of the power converter as a linear system by using a modulator. This approximation simplifies the optimization and allows the calculation of an explicit control law, avoiding the need for online optimization. However, this simplification does not take into account the discrete nature of the power converters.

By including the discrete nature of power converters, it is possible to simplify the optimization problem, allowing its online implementation. Considering the finite number of switching states, and the fast microprocessors available today, calculation of the optimal actuation by online evaluation of each switching state is a real possibility. This consideration allows more flexibility and simplicity in the control scheme, as will be explained in subsequent chapters of this book. As the switching states of the power converters allows finite number of possible actuations, this last approach has been called, in some works, finite control set MPC.

3.3.1 Controller Design

In the design stage of finite control set MPC for the control of a power converter, the following steps are identified:

- Modeling of the power converter identifying all possible switching states and its relation to the input or output voltages or currents.
- Defining a cost function that represents the desired behavior of the system.
- Obtaining discrete-time models that allow one to predict the future behavior of the variables to be controlled.

When modeling a converter, the basic element is the power switch, which can be an IGBT, a thyristor, a gate turn-off thyristor (GTO), or others. The simplest model of this power switches considers an ideal switch with only two states: on and off. Therefore, the total number of switching states of a power converter is equal to the number of different combinations of the two switching states of each switch. However, some combinations are not possible, for example, those combinations that short-circuit the DC link.

As a general rule, the number of possible switching states N is

$$N = x^y \tag{3.5}$$

where x is the number of possible states of each leg of the converter, and y is the number of phases (or legs) of the converter. In this way a three-phase, two-level converter has $N = 2^3 = 8$ possible switching states, a three-phase, three-level converter has $N = 3^3 = 27$ switching states, and a five-phase, two-level converter has $N = 2^5 = 32$ switching states. In some multilevel topologies the number of switching states of the converter can be very high, as in a three-phase, nine-level cascaded H-bridge inverter, where the number of switching states is more than 16 million.

Another aspect of the model of the converter is the relation between the switching states and the voltage levels, in the case of single-phase converters, or voltage vectors, in the case of three-phase or multi-phase converters. For current source converters, the possible switching states are related to current vectors instead of voltage vectors. It can be found that, in several cases, two or more switching states generate the same voltage vector.

For example, in a three-phase, two-level converter, the eight switching states generate seven different voltage vectors, with two switching states generating the zero vector. In a three-phase, three-level converter there is a major redundancy, with 27 switching states generating 19 different voltage vectors. Figure 3.3 depicts the relation between switching

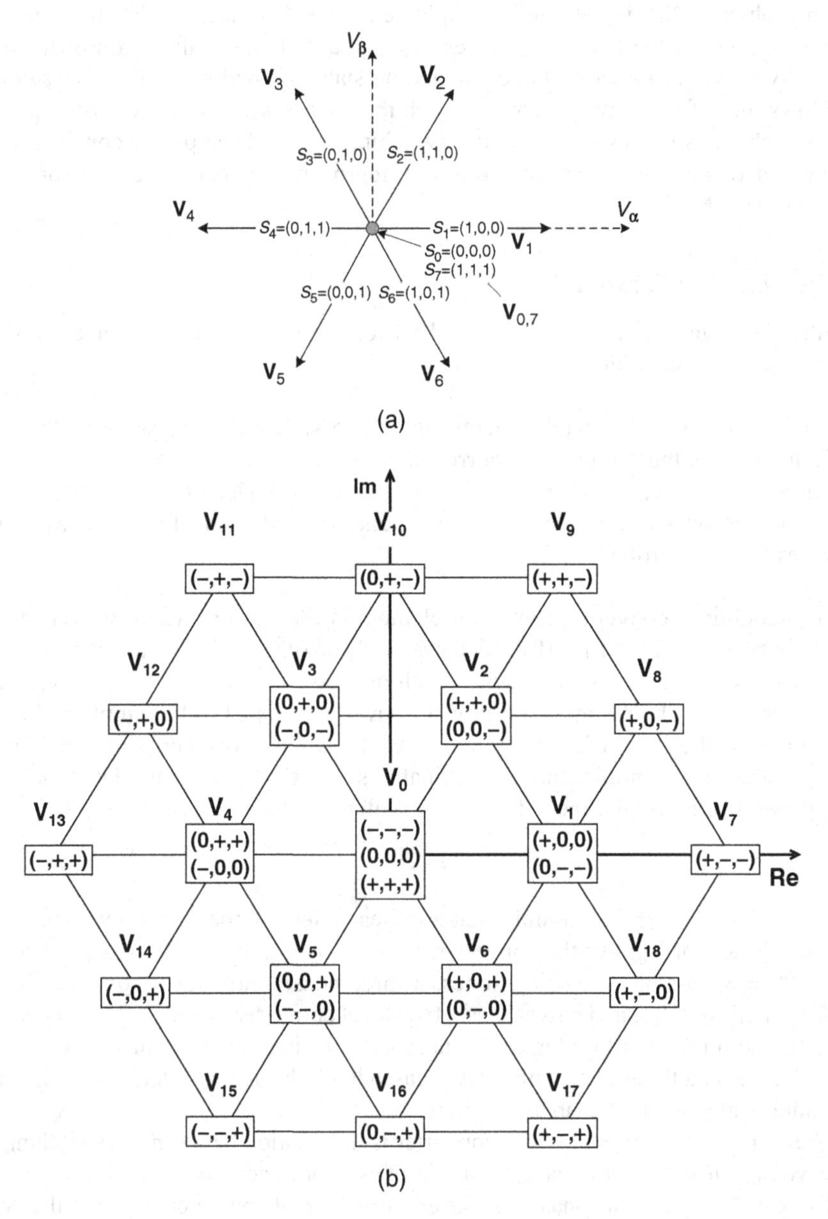

Figure 3.3 Voltage vectors generated by different converters. (a) Three-phase, two-level inverter. (b) Three-phase, three-level inverter

states and voltage vectors for two different converter topologies. In some other topologies, the method of calculating the possible switching states may be different.

Each different application imposes several control requirements on the systems such as current control, torque control, power control, low switching frequency, etc. These requirements can be expressed as a cost function to be minimized. The most basic cost function to be defined is some measure of error between a reference and a predicted variable, for example, load current error, power error, torque error, and others, as will be shown in the following chapters of this book. However, one of the advantages of the predictive control methods is the possibility to control different types of variables and include restrictions on the cost function. In order to deal with the different units and magnitudes of the controlled variables, each term in the cost function is multiplied by a weighting factor that can be used to adjust the importance of each term.

When building the model for prediction, the controlled variables must be considered in order to get discrete-time models that can be used for the prediction of these variables. It is also important to define which variables are measured and which ones are not measured, because in some cases variables that are required for the predictive model are not measured and some kind of estimate will be needed.

To get a discrete-time model it is necessary to use some discretization methods. For first-order systems it is useful, because it is simple, to approximate the derivatives using the Euler forward method, that is, using

$$\frac{dx}{dt} = \frac{x(k+1) - x(k)}{T_s} \tag{3.6}$$

where T_s is the sampling time. However, when the order of the system is higher, the discrete-time model obtained using the Euler method is not so good because the error introduced by this method for higher order systems is significant. For these higher order systems, an exact discretization must be used.

3.3.2 Implementation

When implemented, the controller must consider the following tasks:

- Predict the behavior of the controlled variables for all possible switching states.
- Evaluate the cost function for each prediction.
- Select the switching state that minimizes the cost function.

Implementation of predictive models and a predictive control strategy may encounter different difficulties depending on the type of platform used. When implemented using a fixed-point processor, special attention must be paid to programming in order to get the best accuracy in the fixed-point representation of the variables. On the other hand, when implemented using a floating-point processor, almost the same programming used for simulations can be used in the laboratory.

Depending on the complexity of the controlled system, the number of calculations can be significant and will limit the minimum sampling time. In the simplest case, predictive current control, the calculation time is small, but in other schemes such as torque and flux control, the calculation time is the parameter which determines the allowed sampling time.

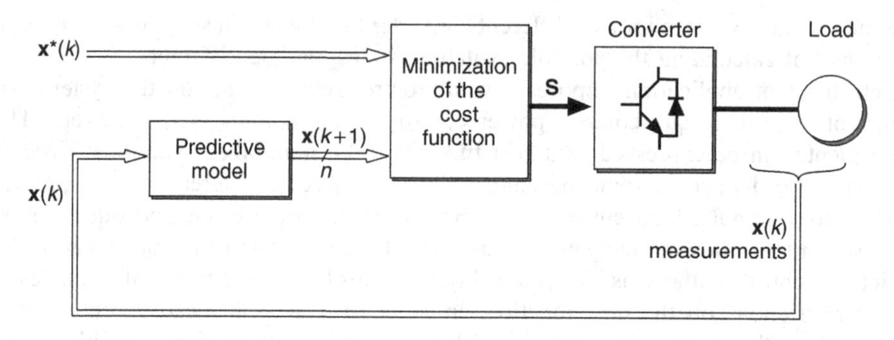

Figure 3.4 General MPC scheme for power converters

To select the switching state which minimizes the cost function, all possible states are evaluated and the optimal valuc is storcd to bc applied next. The number of calculations is directly related to the number of possible switching states. In the case of the three-phase, two-level inverter, to calculate predictions for the eight possible switching states is not a problem, but in the case of multi level and multi-phase systems, a different optimization method must be considered in order to reduce the number of calculations.

3.3.3 General Control Scheme

A general control scheme for MPC applied to power converters and drives is presented in Figure 3.4. The power converter can be from any topology and number of phases, while the generic load shown in the figure can represent an electrical machine, the grid, or any other active or passive load. In this scheme measured variables $\mathbf{x}(k)$ are used in the model to calculate predictions $\mathbf{x}(k+1)$ of the controlled variables for each one of the n possible actuations, that is, switching states, voltages, or currents. Then these predictions are evaluated using a cost function which considers the reference values $\mathbf{x}^*(k)$ and restrictions, and the optimal actuation $\mathbf{S}$ is selected and applied in the converter.

3.4 Summary

This chapter presents an overview of different predictive control methods. The basic principles of MPC and its application for power converters and drives are presented. A general control scheme has been introduced in this chapter and will be considered in all applications included in this book.

References

[1] P. Cortés, M. P. Kazmierkowski, R. M. Kennel, D. E. Quevedo, and J. Rodríguez, "Predictive control in power electronics and drives," *IEEE Transactions on Industrial Electronics*, vol. 55, no. 12, pp. 4312–4324, December 2008.

[2] J. Holtz and S. Stadtfeld, "A predictive controller for the stator current vector of AC machines fed from a switched voltage source," in International Power Electronics Conference, IPEC, Tokyo, pp. 1665–1675, 1983.

[3] P. Mutschler, "A new speed-control method for induction motors," in Conference Record of PCIM'98, Nuremberg, pp. 131–136, May 1998.
[4] T. Kawabata, T. Miyashita, and Y. Yamamoto, "Dead beat control of three phase PWM inverter," *IEEE Transactions on Power Electronics*, vol. 5, no. 1, pp. 21–28, January 1990.
[5] O. Kukrer, "Discrete-time current control of voltage-fed three-phase PWM inverters," *IEEE Transactions on Industrial Electronics*, vol. 11, no. 2, pp. 260–269, March 1996.
[6] S. Kouro, P. Cortés, R. Vargas, U. Ammann, and J. Rodríguez, "Model predictive control – a simple and powerful method to control power converters," *IEEE Transactions on Industrial Electronics*, vol. 56, no. 6, pp. 1826–1838, June 2009.
[7] E. F. Camacho and C. Bordons, *Model Predictive Control* Springer Verlag, 1999.
[8] J. M. Maciejowski, *Predictive Control with Constraints*. Englewood Cliffs, NJ: Prentice Hall, 2002.
[9] G. C. Goodwin, M. M. Seron, and J. A. D. Dona, *Constrained Control and Estimation – An Optimization Perspective*. Springer Verlag, 2005.
[10] C. E. Garcia, D. M. Prett, and M. Morari, "Model predictive control: theory and practice – a survey," *Automatica*, vol. 25, no. 3, pp. 335–348, May 1989.
[11] C. Bordons and E. Camacho, "A generalized predictive controller for a wide class of industrial processes," *IEEE Transactions on Control Systems Technology*, vol. 6, no. 3, pp. 372–387, May, 1998.
[12] D. W. Clarke, C. Mohtadi, and P. S. Tuffs, "Generalized predictive control – part I. The basic algorithm," *Automatica*, vol. 23, no. 2, pp. 137–148, 1987.
[13] E. El-Kholy, "Generalized predictive controller for a boost ac to dc converter fed dc motor," in International Conference on Power Electronics and Drives Systems 2005. PEDS 2005, vol. 2, pp. 1090–1095, November 2005.
[14] S. Effler, A. Kelly, M. Halton, and K. Rinne, "Automated optimization of generalized model predictive control for dc-dc converters," in IEEE Power Electronics Specialists Conference 2008. PESC 2008, pp. 134–139, June 2008.
[15] K. Low, "A digital control technique for a single-phase pwm inverter," *IEEE Transactions on Industrial Electronics*, vol. 45, no. 4, pp. 672–674, August 1998.
[16] R. Kennel, A. Linder, and M. Linke, "Generalized predictive control (GPC): ready for use in drive applications?" in IEEE 32nd Annual Power Electronics Specialists Conference, 2001 PESC, vol. 4, pp. 1839–1844, 2001.
[17] P. Egiguren, O. Caramazana, A. Garrido Hernandez, and I. Garrido Hernandez, "SVPWM linear generalized predictive control of induction motor drives," in IEEE International Symposium on Industrial Electronics 2008. ISIE 2008, pp. 588–593, June 2008.
[18] S. Hassaine, S. Moreau, C. Ogab, and B. Mazari, "Robust speed control of PMSM using generalized predictive and direct torque control techniques," in IEEE International Symposium on Industrial Electronics 2007. ISIE 2007, pp. 1213–1218, June 2007.
[19] A. Beccuti, S. Mariethoz, S. Cliquennois, S. Wang, and M. Morari, "Explicit model predictive control of dc-dc switched-mode power supplies with extended Kalman filtering," *IEEE Transactions on Industrial Electronics*, vol. 56, no. 6, pp. 1864–1874, June 2009.
[20] S. Mariethoz and M. Morari, "Explicit model-predictive control of a PWM inverter with an LCL filter," *IEEE Transactions on Industrial Electronics*, vol. 56, no. 2, pp. 389–399, February 2009.
[21] S. Mariethoz, A. Domahidi, and M. Morari, "Sensorless explicit model predictive control of permanent magnet synchronous motors," in IEEE International Electric Machines and Drives Conference 2009. IEMDC '09, pp. 1250–1257, May 2009.

Part Two

Model Predictive Control Applied to Power Converters

4

Predictive Control of a Three-Phase Inverter

4.1 Introduction

Current control is one of the most studied problems in power electronics [1–3], so it is very important to study as a first step the application of MPC in a current control scheme. In addition, the three-phase, two-level inverter is a very well-known topology that can be found in most drive applications.

This chapter presents a MPC scheme for current control in a three-phase inverter, which is based on the control scheme reported in [4]. The control scheme and working principle will be explained in more detail than other applications in this book because the same ideas and explanations can be extended for all examples presented in the following chapters.

4.2 Predictive Current Control

The proposed predictive control strategy is based on the fact that only a finite number of possible switching states can be generated by a static power converter and that models of the system can be used to predict the behavior of the variables for each switching state. For the selection of the appropriate switching state to be applied, a selection criterion must be defined. This criterion consists of a cost function that will be evaluated for the predicted values of the variables to be controlled. Prediction of the future value of these variables is calculated for each possible switching state and then the state that minimizes the cost function is selected.

This control strategy can be summarized in the following steps:

- Define a cost function g.
- Build a model of the converter and its possible switching states.
- Build a model of the load for prediction.

Predictive Control of Power Converters and Electrical Drives, First Edition. Jose Rodriguez and Patricio Cortes.

A discrete-time model of the load is needed to predict the behavior of the variables evaluated by the cost function, that is, the load currents.

4.3 Cost Function

The objective of the current control scheme is to minimize the error between the measured currents and the reference values. This requirement can be written in the form of a cost function. The cost function is expressed in orthogonal coordinates and measures the error between the references and the predicted currents:

$$g = |i_\alpha^*(k+1) - i_\alpha^{\mathrm{p}}(k+1)| + |i_\beta^*(k+1) - i_\beta^{\mathrm{p}}(k+1)| \tag{4.1}$$

where $i_\alpha^{\mathrm{p}}(k+1)$ and $i_\beta^{\mathrm{p}}(k+1)$ are the real and imaginary parts of the predicted load current vector $\mathbf{i}^{\mathrm{p}}(k+1)$, for a given voltage vector. This prediction is obtained using the load model, which will be explained in detail in the sections below. The reference currents $i_\alpha^*(k+1)$ and $i_\beta^*(k+1)$ are the real and imaginary parts of the reference current vector $\mathbf{i}^*(k+1)$. For simplicity, we will assume that this reference current does not change sufficiently in one sampling interval, so we will consider $\mathbf{i}^*(k+1) = \mathbf{i}^*(k)$. This assumption may introduce a one sample delay in the reference tracking, which is not a problem if a high sampling frequency is considered. In other cases it is possible to extrapolate the future reference value, as will be explained in Chapter 12. This reference is generated from an external control loop, for example, field-oriented control of an induction machine.

A block diagram of the predictive control strategy applied to the current control for a three-phase inverter is shown in Figure 4.1. The current control is performed in four steps, as described in Table 4.1.

4.4 Converter Model

The power circuit of the three-phase inverter converts electrical power from DC to AC form using the electrical scheme shown in Figure 4.2. Considering that the two switches in each inverter phase operate in a complementary mode in order to avoid short-circuiting

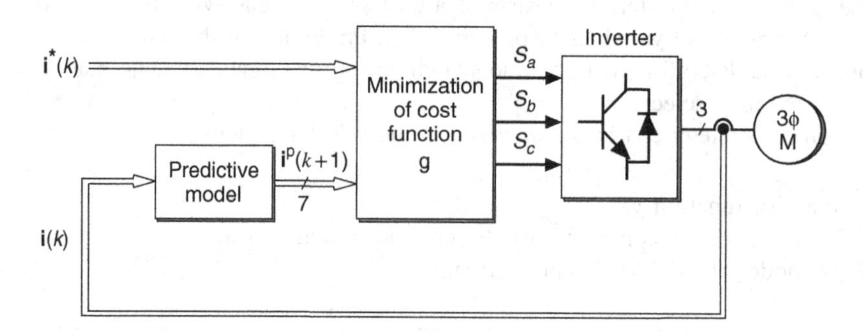

Figure 4.1 Predictive current control block diagram (Rodriguez *et al.*, 2007 © IEEE)

Table 4.1 Predictive current control algorithm

1. The value of the reference current $\mathbf{i}^*(k)$ is obtained (from an outer control loop), and the load current $\mathbf{i}(k)$ is measured.
2. The model of the system is used to predict the value of the load current in the next sampling interval $\mathbf{i}(k+1)$ for each of the different voltage vectors.
3. The cost function g evaluates the error between the reference and predicted currents in the next sampling interval for each voltage vector.
4. The voltage that minimizes the current error is selected and the corresponding switching state signals are generated.

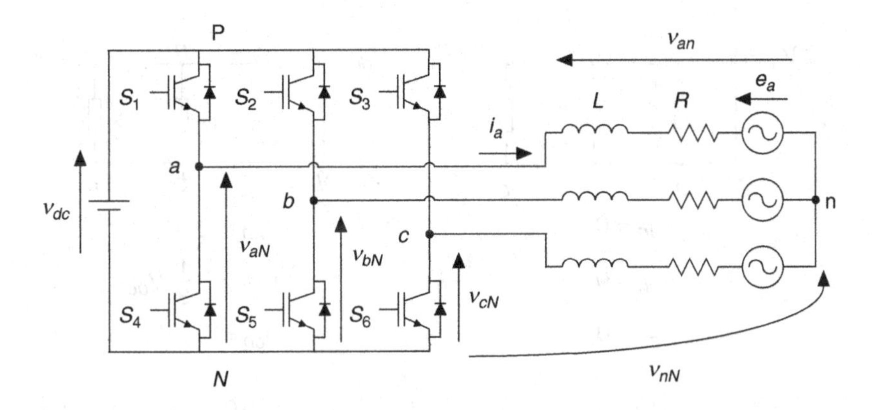

Figure 4.2 Voltage source inverter power circuit

the DC source, the switching state of the power switches S_x, with $x = 1, \ldots, 6$, can be represented by the switching signals S_a, S_b, and S_c defined as follows:

$$S_a = \begin{cases} 1 \text{ if } S_1 \text{ on and } S_4 \text{ off} \\ 0 \text{ if } S_1 \text{ off and } S_4 \text{ on} \end{cases} \tag{4.2}$$

$$S_b = \begin{cases} 1 \text{ if } S_2 \text{ on and } S_5 \text{ off} \\ 0 \text{ if } S_2 \text{ off and } S_5 \text{ on} \end{cases} \tag{4.3}$$

$$S_c = \begin{cases} 1 \text{ if } S_3 \text{ on and } S_6 \text{ off} \\ 0 \text{ if } S_3 \text{ off and } S_6 \text{ on} \end{cases} \tag{4.4}$$

These switching signals define the value of the output voltages

$$v_{aN} = S_a V_{dc} \tag{4.5}$$

$$v_{bN} = S_b V_{dc} \tag{4.6}$$

$$v_{cN} = S_c V_{dc} \tag{4.7}$$

where V_{dc} is the DC source voltage.

Considering the unitary vector $\mathbf{a} = e^{j2\pi/3} = -\frac{1}{2} + j\sqrt{3}/2$, which represents the 120° phase displacement between the phases, the output voltage vector can be defined as

$$\mathbf{v} = \frac{2}{3}(v_{aN} + \mathbf{a}v_{bN} + \mathbf{a}^2 v_{cN}) \tag{4.8}$$

where v_{aN}, v_{bN}, and v_{cN} are the phase-to-neutral (N) voltages of the inverter.

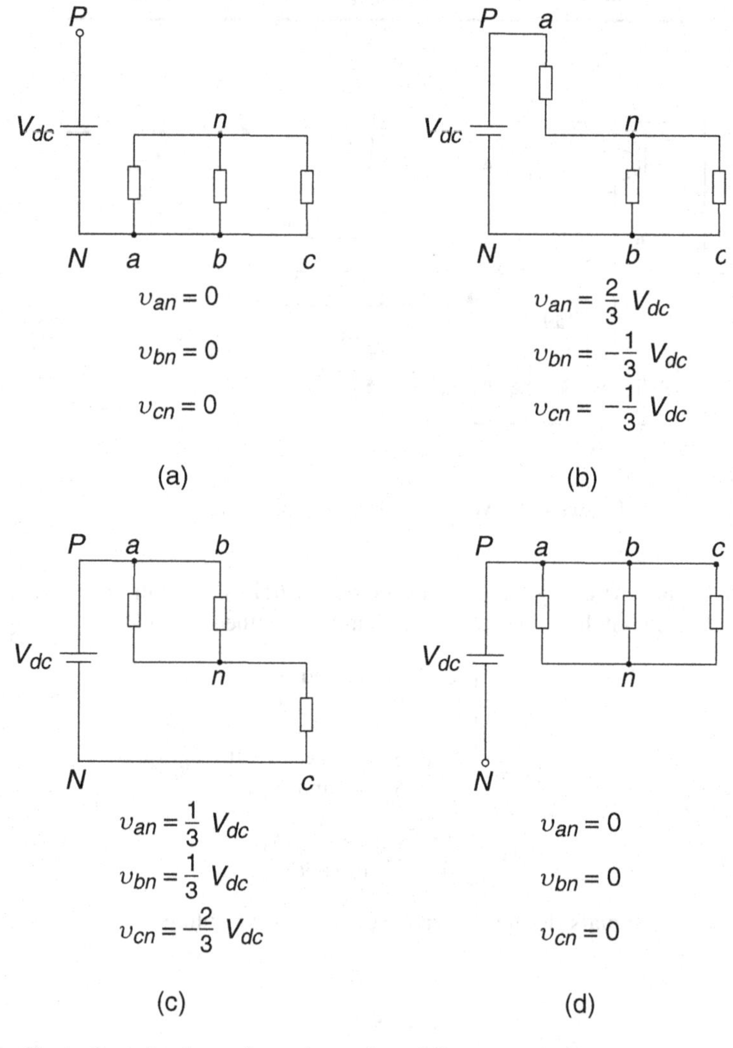

Figure 4.3 Equivalent load configurations for different switching states. (a) Switching state $(0, 0, 0)$ (voltage vector $\mathbf{V}_0$). (b) Switching state $(1, 0, 0)$ (voltage vector $\mathbf{V}_1$). (c) Switching state $(1, 1, 0)$ (voltage vector $\mathbf{V}_2$). (d) Switching state $(1, 1, 1)$ (voltage vector $\mathbf{V}_7$)

In this way, switching state $(S_a, S_b, S_c) = (0, 0, 0)$ generates voltage vector $\mathbf{V}_0$ defined as

$$\mathbf{V}_0 = \frac{2}{3}(0 + \mathbf{a}0 + \mathbf{a}^2 0) = 0 \tag{4.9}$$

and corresponds to the circuit shown in Figure 4.3a.

Switching state $(1, 0, 0)$ generates voltage vector $\mathbf{V}_1$ defined as

$$\mathbf{V}_1 = \frac{2}{3}(V_{dc} + \mathbf{a}0 + \mathbf{a}^2 0) = \frac{2}{3}V_{dc} \tag{4.10}$$

Voltage vector $\mathbf{V}_2$ is generated by switching state $(1, 1, 0)$ and is defined as

$$\begin{aligned}\mathbf{V}_2 &= \frac{2}{3}(V_{dc} + \mathbf{a}V_{dc} + \mathbf{a}^2 0) \\ &= \frac{2}{3}\left(V_{dc} + \left(-\frac{1}{2} + j\frac{\sqrt{3}}{2}\right)V_{dc}\right) = \frac{V_{dc}}{3} + j\frac{\sqrt{3}}{3}V_{dc}\end{aligned} \tag{4.11}$$

and corresponds to the circuit shown in Figure 4.3b.

Switching state $(1, 1, 1)$ generates voltage vector $\mathbf{V}_7$ that is calculated as

$$\mathbf{V}_7 = \frac{2}{3}(V_{dc} + \mathbf{a}V_{dc} + \mathbf{a}^2 V_{dc}) = \frac{2}{3}V_{dc}(1 + \mathbf{a} + \mathbf{a}^2) = 0 \tag{4.12}$$

Different switching states will generate different configurations of the three-phase load connected to the DC source, as shown in Figure 4.3.

Considering all the possible combinations of the gating signals S_a, S_b, and S_c, eight switching states and consequently eight voltage vectors are obtained, as shown in Table 4.2. In Figure 4.4 note that $\mathbf{V}_0 = \mathbf{V}_7$, resulting in a finite set of only seven different voltage vectors in the complex plane.

Taking into account modulation techniques, like PWM, the inverter can be approximated as a linear system. Nevertheless, throughout this chapter the inverter will be considered as a non linear discrete system with only seven different states as possible outputs.

Table 4.2 Switching states and voltage vectors

S_a	S_b	S_c	Voltage vector $\mathbf{V}$
0	0	0	$\mathbf{V}_0 = 0$
1	0	0	$\mathbf{V}_1 = \frac{2}{3}V_{dc}$
1	1	0	$\mathbf{V}_2 = \frac{1}{3}V_{dc} + j\frac{\sqrt{3}}{3}V_{dc}$
0	1	0	$\mathbf{V}_3 = -\frac{1}{3}V_{dc} + j\frac{\sqrt{3}}{3}V_{dc}$
0	1	1	$\mathbf{V}_4 = -\frac{2}{3}V_{dc}$
0	0	1	$\mathbf{V}_5 = -\frac{1}{3}V_{dc} - j\frac{\sqrt{3}}{3}V_{dc}$
1	0	1	$\mathbf{V}_6 = \frac{1}{3}V_{dc} - j\frac{\sqrt{3}}{3}V_{dc}$
1	1	1	$\mathbf{V}_7 = 0$

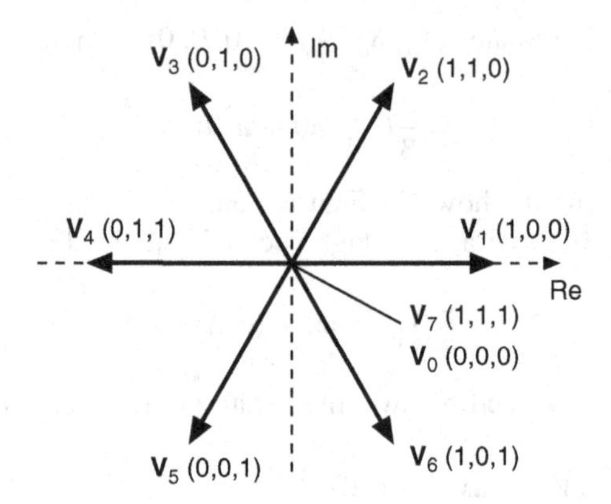

Figure 4.4 Voltage vectors in the complex plane

We should note that a more complex model of the converter model could be used for higher switching frequencies. It might include modeling dead time, IGBT saturation voltage, and diode forward voltage drop, for example. However, in this book, emphasis has been put on simplicity, so a simple model of the inverter will be used.

4.5 Load Model

Taking into account the definitions of variables from the circuit shown in Figure 4.2, the equations for load current dynamics for each phase can be written as

$$v_{aN} = L\frac{di_a}{dt} + Ri_a + e_a + v_{nN} \tag{4.13}$$

$$v_{bN} = L\frac{di_b}{dt} + Ri_b + e_b + v_{nN} \tag{4.14}$$

$$v_{cN} = L\frac{di_c}{dt} + Ri_c + e_c + v_{nN} \tag{4.15}$$

where R is the load resistance and L the load inductance.

By substituting (4.13)–(4.15) into (4.8) a vector equation for the load current dynamics can be obtained:

$$\mathbf{v} = L\frac{d}{dt}\left(\frac{2}{3}(i_a + \mathbf{a}i_b + \mathbf{a}^2 i_c)\right) + R\left(\frac{2}{3}(i_a + \mathbf{a}i_b + \mathbf{a}^2 i_c)\right) + \frac{2}{3}(e_a + \mathbf{a}e_b + \mathbf{a}^2 e_c) + \frac{2}{3}(v_{nN} + \mathbf{a}v_{nN} + \mathbf{a}^2 v_{nN}) \tag{4.16}$$

Considering the space vector definition for the inverter voltage given by (4.8), and the following definitions for load current and back-emf space vectors

$$\mathbf{i} = \frac{2}{3}(i_a + \mathbf{a}i_b + \mathbf{a}^2 i_c) \tag{4.17}$$

$$\mathbf{e} = \frac{2}{3}(e_a + \mathbf{a}e_b + \mathbf{a}^2 e_c) \tag{4.18}$$

and assuming the last term of (4.16) equal to zero

$$\frac{2}{3}(v_{nN} + \mathbf{a}v_{nN} + \mathbf{a}^2 v_{nN}) = v_{nN}\frac{2}{3}(1 + \mathbf{a} + \mathbf{a}^2) = 0 \tag{4.19}$$

then the load current dynamics can be described by the vector differential equation

$$\mathbf{v} = R\mathbf{i} + L\frac{d\mathbf{i}}{dt} + \mathbf{e} \tag{4.20}$$

where $\mathbf{v}$ is the voltage vector generated by the inverter, $\mathbf{i}$ is the load current vector, and $\mathbf{e}$ the load back-emf vector.

Note that for simulation and experimental results, the load back-emf is assumed to be sinusoidal with constant amplitude and constant frequency.

4.6 Discrete-Time Model for Prediction

This section describes the discretization process of the load current equation (4.20) for a sampling time T_s. The discrete-time model will be used to predict the future value of load current from voltages and measured currents at the kth sampling instant. Several discretization methods can be used in order to obtain a discrete-time model suitable for the calculation of predictions. Considering that the load can be modeled as a first-order system, the discrete-time model can be obtained by a simple approximation of the derivative. However, for more complex systems this approximation may introduce errors into the model and a more accurate discretization method is required.

The load current derivative $d\mathbf{i}/dt$ is replaced by a forward Euler approximation. That is, the derivative is approximated as follows:

$$\frac{d\mathbf{i}}{dt} \approx \frac{\mathbf{i}(k+1) - \mathbf{i}(k)}{T_s} \tag{4.21}$$

which is substituted in (4.20) to obtain an expression that allows prediction of the future load current at time $k+1$, for each one of the seven values of voltage vector $\mathbf{v}(k)$ generated by the inverter. This expression is

$$\mathbf{i}^{\mathbf{p}}(k+1) = \left(1 - \frac{RT_s}{L}\right)\mathbf{i}(k) + \frac{T_s}{L}\left(\mathbf{v}(k) - \hat{\mathbf{e}}(k)\right) \tag{4.22}$$

where $\hat{\mathbf{e}}(k)$ denotes the estimated back-emf. The superscript $\mathbf{p}$ denotes the predicted variables.

The back-emf can be calculated from (4.20) considering measurements of the load voltage and current with the following expression;

$$\hat{\mathbf{e}}(k-1) = \mathbf{v}(k-1) - \frac{L}{T_s}\mathbf{i}(k) - \left(R - \frac{L}{T_s}\right)\mathbf{i}(k-1) \tag{4.23}$$

where $\hat{\mathbf{e}}(k-1)$ is the estimated value of $\mathbf{e}(k-1)$. The present back-emf $\mathbf{e}(k)$, needed in (12.2), can be estimated using an extrapolation of the past values of the estimated back-emf. Alternatively, as the frequency of the back-emf is much less than the sampling frequency, we will suppose that it does not change considerably in one sampling interval and, thus, assume $\hat{\mathbf{e}}(k) = \hat{\mathbf{e}}(k-1)$.

4.7 Working Principle

In order to illustrate how the predictive control strategy works, a detailed example is shown in Figure 4.5 and Figure 4.6. Here, load currents i_α, i_β, and their references are shown for a complete period of the reference. Using the measurement $\mathbf{i}(k)$ and all switching states of the voltage vector $\mathbf{v}(k)$, the future currents $\mathbf{i}(k+1)$ are estimated, $\mathbf{i}^{\mathrm{p}}(k+1)$.

In the vectorial plot, shown in Figure 4.5, it can be observed that vector $\mathbf{V}_2$ takes the predicted current vector closest to the reference vector.

As shown in Figure 4.6, current $i_\alpha^{\mathrm{p}}(\mathbf{V}_{0,7})$ corresponds to the predicted current if the voltage vector $\mathbf{V}_0$ or $\mathbf{V}_7$ is applied at time k. It can be seen in this figure that vectors $\mathbf{V}_2$ and $\mathbf{V}_6$ are the ones that minimize the error in the i_α current, and vectors $\mathbf{V}_2$ and $\mathbf{V}_3$ are the ones that minimize the error in the i_β current, so the voltage vector that minimizes the cost function g is $\mathbf{V}_2$.

These figures illustrate the meaning of the cost function as a measure of error or distance between reference and predicted vectors. It is easy to view these errors and distances for the case of current control, but these plots become difficult or impossible to build for more complex cost functions.

From a numerical point of view, the selection of the optimum voltage vector is performed as presented in Figure 4.7. Each voltage vector generates a predicted current that gives a value of the cost function, as listed in the table. It can be observed that, for this example, vector $\mathbf{V}_2$ produces the lowest value of the cost function g. Then, voltage vector $\mathbf{V}_2$ is selected and applied in the inverter.

4.8 Implementation of the Predictive Control Strategy

A flow diagram of the different tasks performed by the predictive controller is shown in Figure 4.8. Here, the outer loop is executed every sampling time, and the inner loop is executed for each possible state, obtaining the optimal switching state to be applied during the next sampling period.

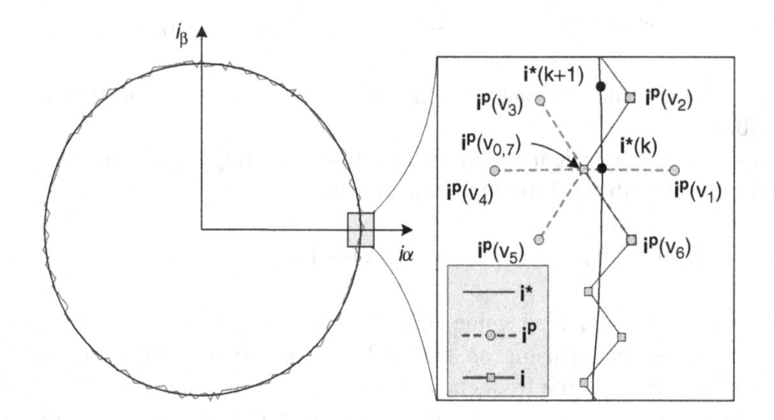

Figure 4.5 Working principle: vectorial plot of the reference and predicted currents

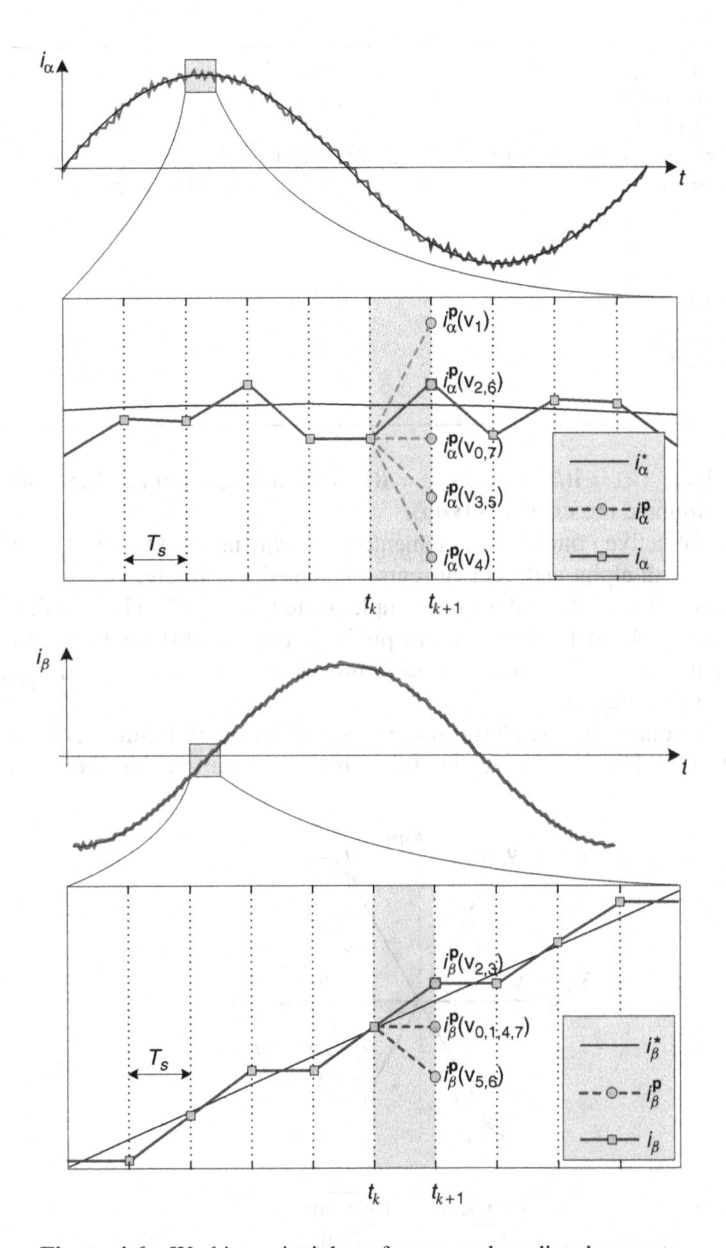

Figure 4.6 Working principle: reference and predicted currents

The timing of the different tasks is shown in Figure 4.9, and, as illustrated here, the most time-consuming task is the prediction and selection of the optimal switching state. This is due to the calculation of the load model and cost function, which is executed seven times, once for each different voltage vector.

The predictive current control strategy is implemented for simulation in MATLAB as a S-Function, containing the following code:

```
ek = v(xop_1) - L/Ts*ik - (R-L/Ts)*ik_1;
g_opt = inf;
for i=1:7
    ik1 = (1-R*Ts/L)*ik + Ts/L*(v(i)-ek);
    g = abs(real(ik_ref-ik1)) + abs(imag(ik_ref-ik1));
    if (g<g_opt)
        g_opt = g;
        x_opt = i;
    end
end
xop_1=xop;
xop=x_opt;
```

where `ik`= $\mathbf{i}(k)$, `ik1`= $\mathbf{i}(k+1)$, `ik_1`= $\mathbf{i}(k-1)$, and `ek`= $\mathbf{e}(k)$. The optimal voltage vector that minimize the error is $\mathbf{v}(\mathbf{xop})$.

When the predictive control is implemented experimentally, the same code is rewritten in C language with alpha and beta currents calculated separately.

Results using the control algorithm implemented in MATLAB/Simulink are shown next, considering (4.22) for load current prediction and (4.23) for back-emf estimation. The system parameters $V_{dc} = 520\,\text{V}$, $L = 10\,\text{mH}$, $R = 10\,\Omega$, and $e = 100\,\text{V}_{peak}$ have been considered for simulations.

Current and voltage in one phase of the load are shown in Figure 4.10 for a sampling time $T_s = 100\,\mu\text{s}$. There is no steady state error in the current but there is a noticeable

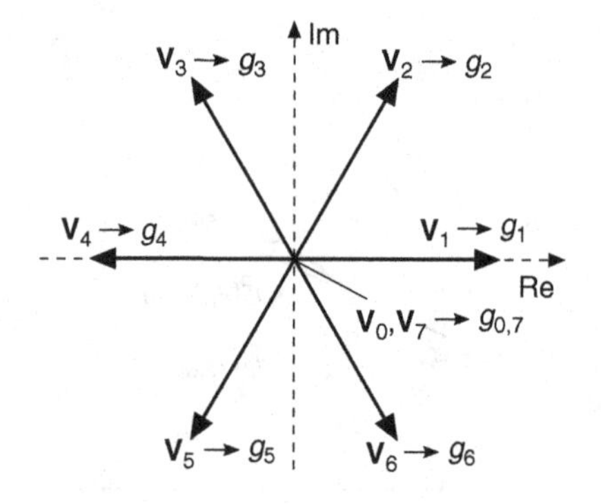

Voltage vector	Cost function	
V_0, V_7	$g_{0,7} = 0.60$	
V_1	$g_1 = 0.82$	
V_2	$g_2 = 0.24$	←optimum
V_3	$g_3 = 0.42$	
V_4	$g_4 = 0.96$	
V_5	$g_5 = 1.24$	
V_6	$g_6 = 1.19$	

Figure 4.7 Working principle: values of the cost function for each voltage vector

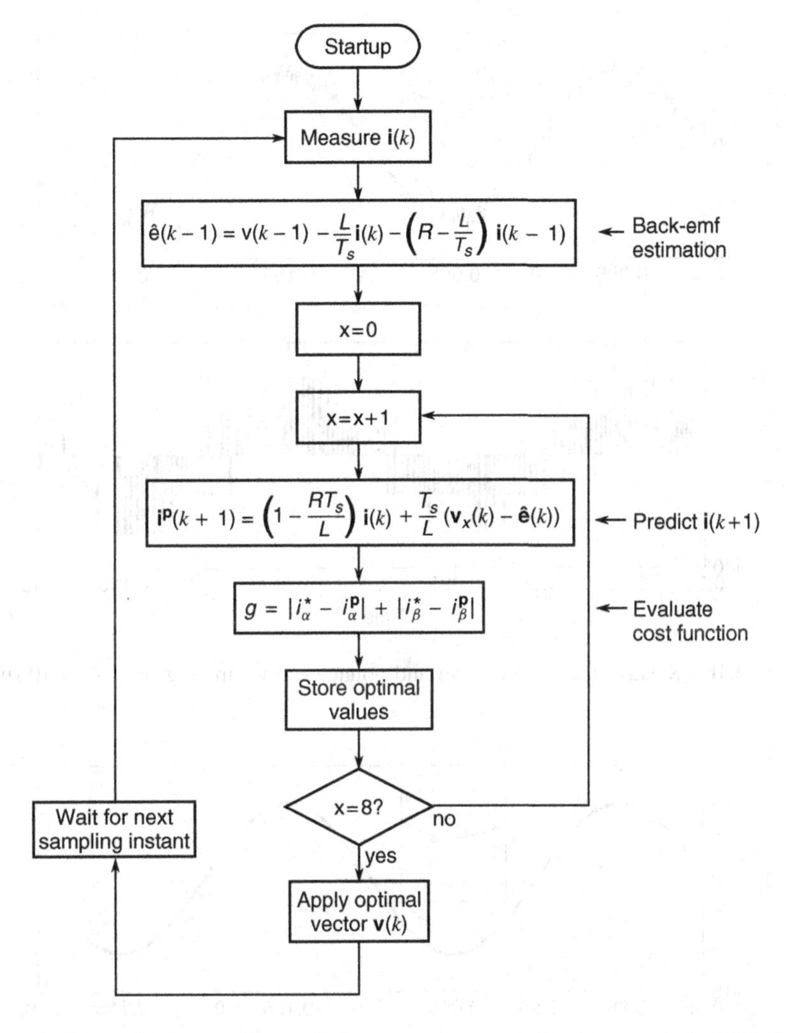

Figure 4.8 Flow diagram of the predictive current control (Rodriguez *et al.*, 2007 © IEEE)

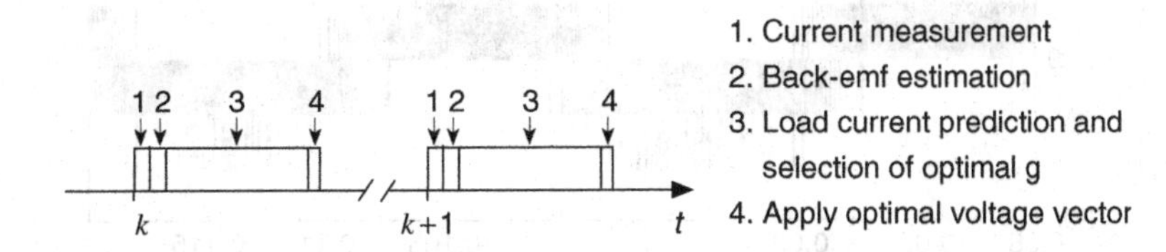

Figure 4.9 Timing of the different tasks (Rodriguez *et al.*, 2007 © IEEE)

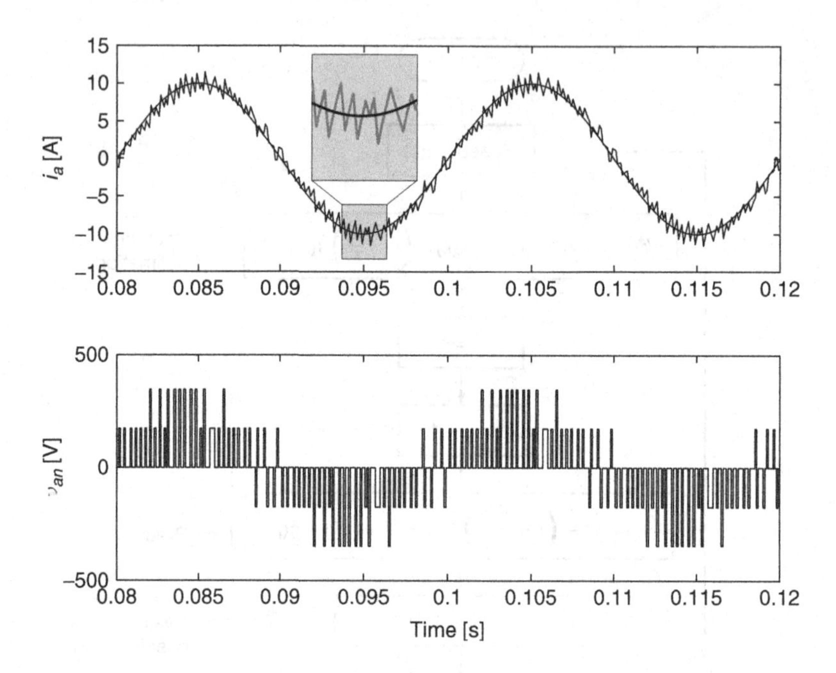

Figure 4.10 Steady state load current and voltage for a sampling time $T_s = 100\,\mu s$

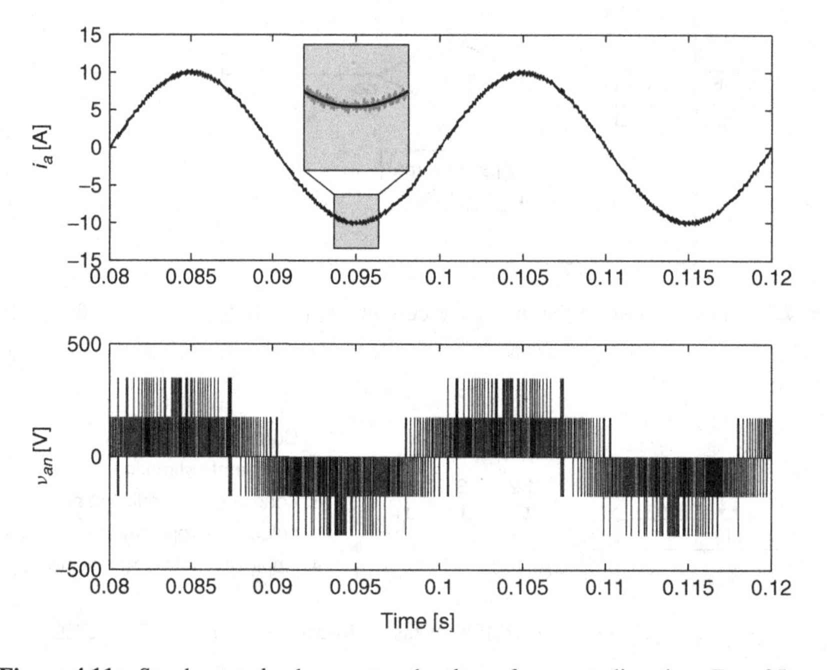

Figure 4.11 Steady state load current and voltage for a sampling time $T_s = 25\,\mu s$

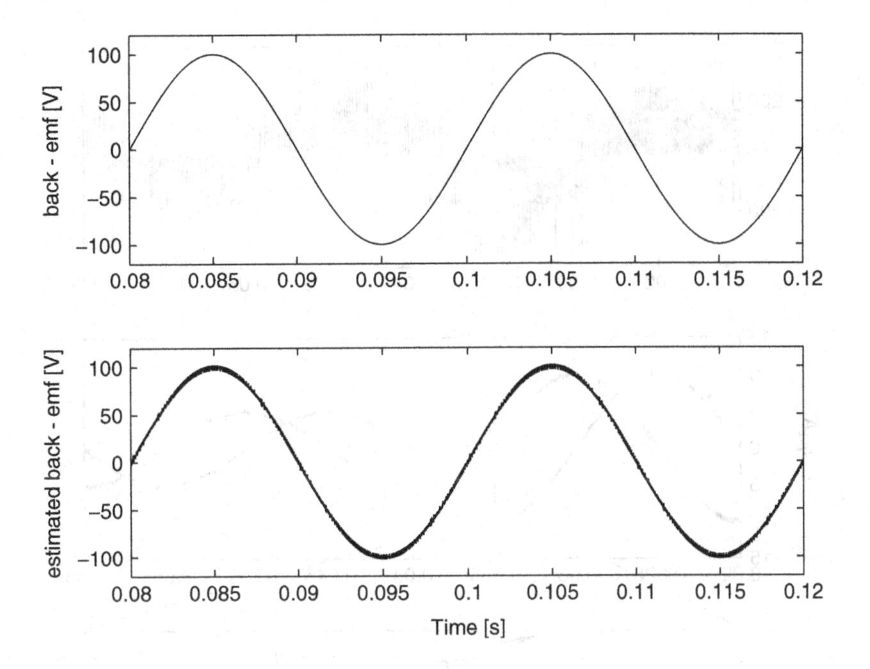

Figure 4.12 Real and estimated back-emf

ripple. This ripple is reduced considerably when a smaller sampling time is used, as shown in Figure 4.11 for a sampling time $T_s = 25\,\mu s$. However, by reducing the sampling time, the switching frequency is increased as can be seen by comparing the load voltages in Figure 4.10 and Figure 4.11.

Real and estimated back-emf are shown in Figure 4.12. Back-emf is estimated using (4.23) with a sampling time $T_s = 25\,\mu s$.

The response of the system to a step in the amplitude of the reference current vector $\mathbf{i}^*$ is shown in Figure 4.13. It can be observed that the load currents follow their references with fast dynamics. It is also shown in this figure how the load voltage changes during the reference current step.

The predictive current control strategy was implemented with the experimental setup described in Figure 4.14. A Danfoss VLT5008 5.5 kW three-phase inverter with an *RL* load is used. The DC link is fed by a three-phase diode bridge rectifier. The inverter is controlled externally through an interface and protection card (IPC). A TMS320C6713 floating-point DSP was used for the control. The gate drive signals for each inverter leg are sent from the DSP to the IPC through fiber optic cables. Current measurements of two phases are sent back from the inverter to the DSP through coaxial cables. A FPGA-based daughter card handles the analog to digital conversion, digital to analog conversion, and the digital outputs for the DSP. The execution time for the implemented current control algorithm was about 7 μs. The parameters of the experimental setup are $V_{dc} = 520\,\mathrm{V}$, $L = 10\,\mathrm{mH}$, $R = 10\,\Omega$, $e = 0\,\mathrm{V}_{peak}$, and $T_s = 25\,\mu s$.

The behavior of the load currents and the voltage in one phase at steady state operation is shown in Figure 4.15. The amplitude of the reference current was set to 10 A with a frequency of 50 Hz. Currents are sinusoidal with low harmonic distortion.

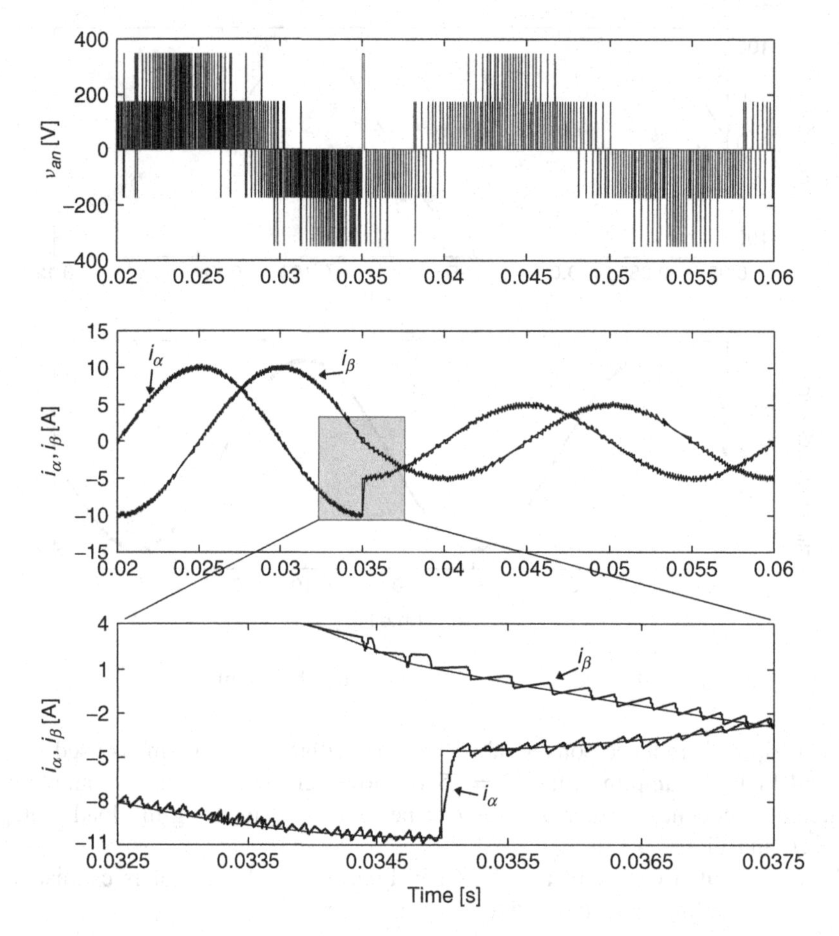

Figure 4.13 Load currents for a step in the amplitude of the reference current vector $\mathbf{i}^*$, with sampling time $T_s = 25\,\mu\text{s}$

Results for a step change in the amplitude of the reference i_α^* from 5 to 10 A are shown in Figure 4.16. The amplitude of the reference current i_β^* is kept at 10 A. It can be observed that the load current i_α reaches its reference with very fast dynamics while i_β is not affected by this step change. This result shows that the currents are decoupled in the predictive current control scheme. The load voltage v_{an} for the same test is shown in Figure 4.17. It can be seen that during the step change of current i_α the load voltage is kept at its maximum value until the reference current is achieved.

The behavior of the load currents and load voltage for a step in the amplitude of the reference current vector is shown in Figure 4.18. The amplitude of the reference current changes from 10 to 5 A and later from 5 to 10 A for all phases. The reference current i_a^* is also shown in the figure. It can be seen that the amplitude of the load currents changes very quickly following the change of the reference. It is also shown in the figure how the load voltage v_{an} changes during this test.

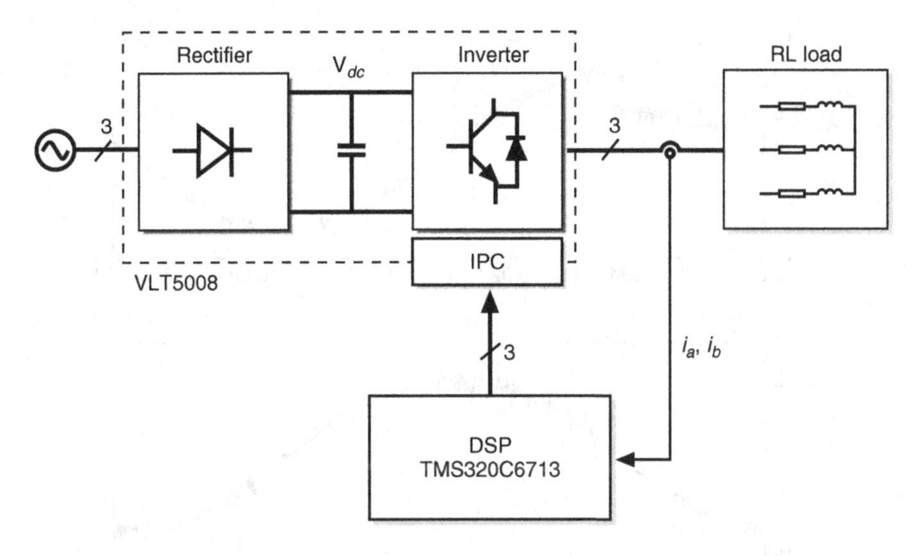

Figure 4.14 Overview of experimental system setup

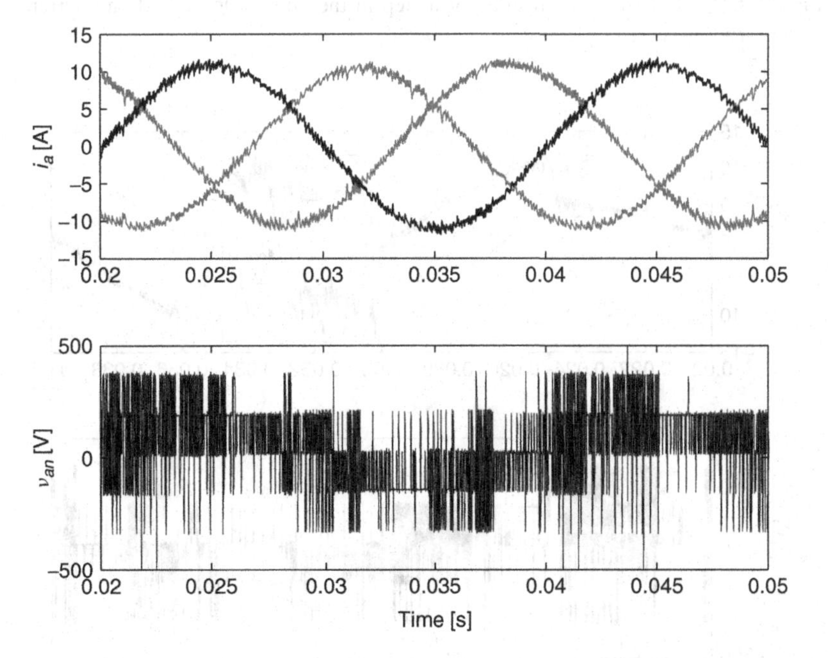

Figure 4.15 Experimental results in steady state. Load currents and voltage in one phase of the load

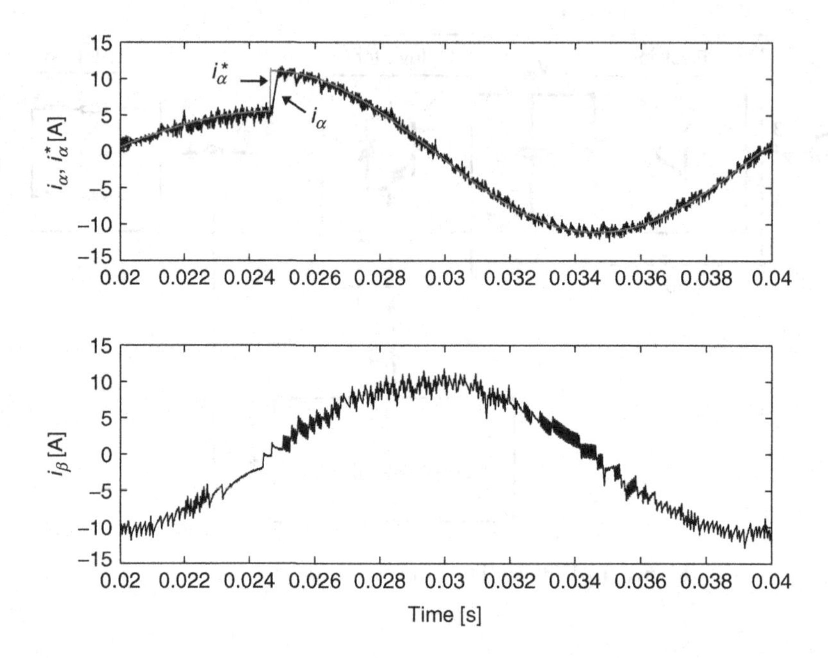

Figure 4.16 Experimental results for a step in the amplitude of i_α^*. Load currents

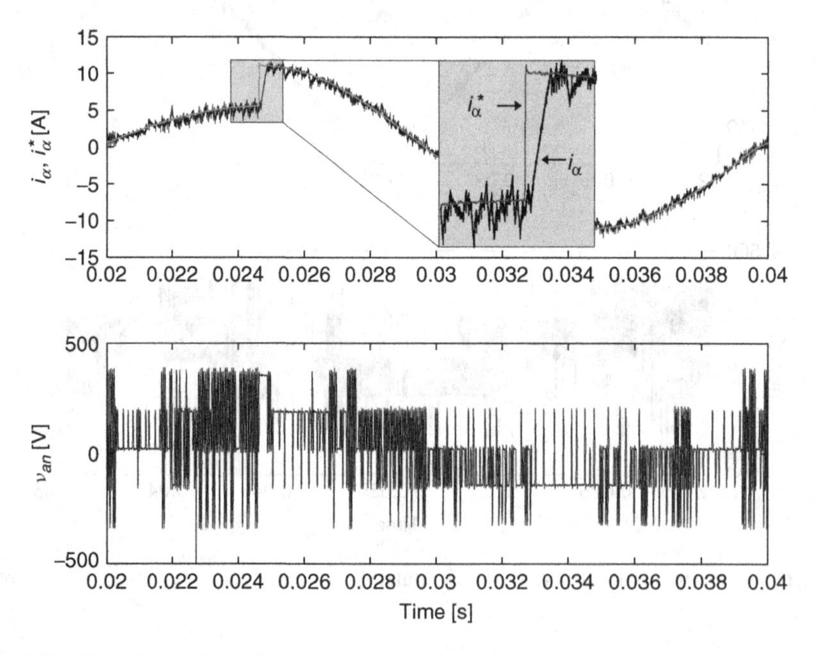

Figure 4.17 Experimental results for a step in the amplitude of i_α^*. Load current and voltage

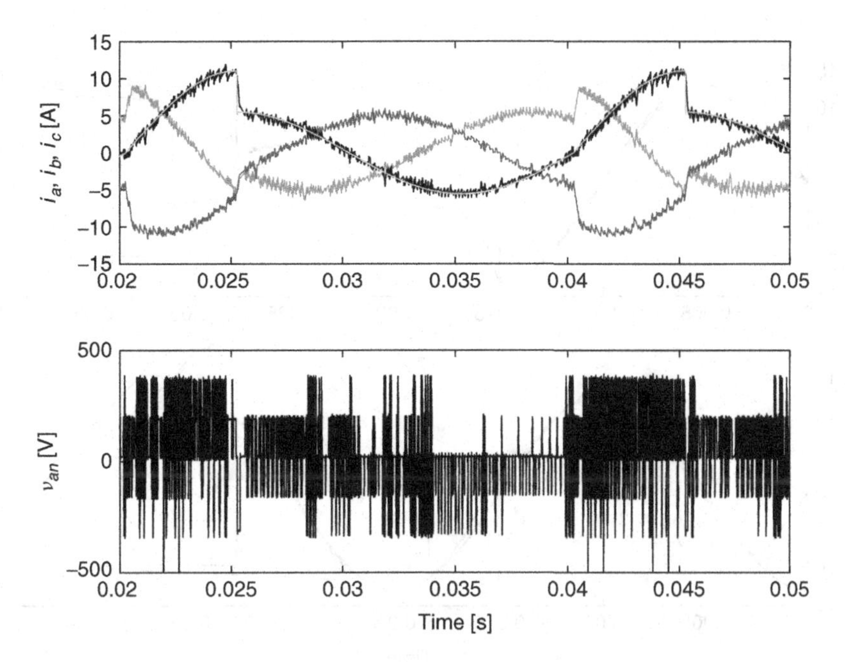

Figure 4.18 Experimental results for a step in the amplitude of the load current reference. Load currents and voltage in one phase

4.9 Comparison to a Classical Control Scheme

A comparison of the predictive current control to classical control with PWM is presented in Figure 4.19a and Figure 4.19b. Here the amplitude of reference current i_α^* is reduced from 13 to 5.2 A at instant $t = 0.015$ s while keeping the amplitude of current i_β^* fixed. This is done to assess the decoupling capability of the current control loop. The load currents obtained using PI controllers with PWM, shown in Figure 4.19a, present noticeable coupling between i_α and i_β and a slower response due to the dynamics of the closed current loops. The response of the predictive current control, for the same test, is shown in Figure 4.19b. Its dynamic response is fast with an inherent decoupling between both current components.

In addition to the reference tracking capabilities of any current control method, another important performance measure is the output voltage spectra generated by the inverter. The voltage spectra for the two control methods are compared in Figure 4.20.

The frequency spectrum of Figure 4.20a shows that the harmonic content, generated when using classical current control, is concentrated around the carrier frequency due to the PWM. Finally, Figure 4.20b presents the frequency spectrum obtained with predictive current control. The voltage spectrum of the predictive control method is characterized by discrete spectral lines similar to those of classical current control, although these spectral lines are more spread over the frequency range. Although the switching frequency for predictive control is variable, it is limited. In fact, the switching state of the inverter can be changed only once during each sampling instant, thus switching frequency is limited to half the sampling frequency (f_s). However, switching states do not change in every

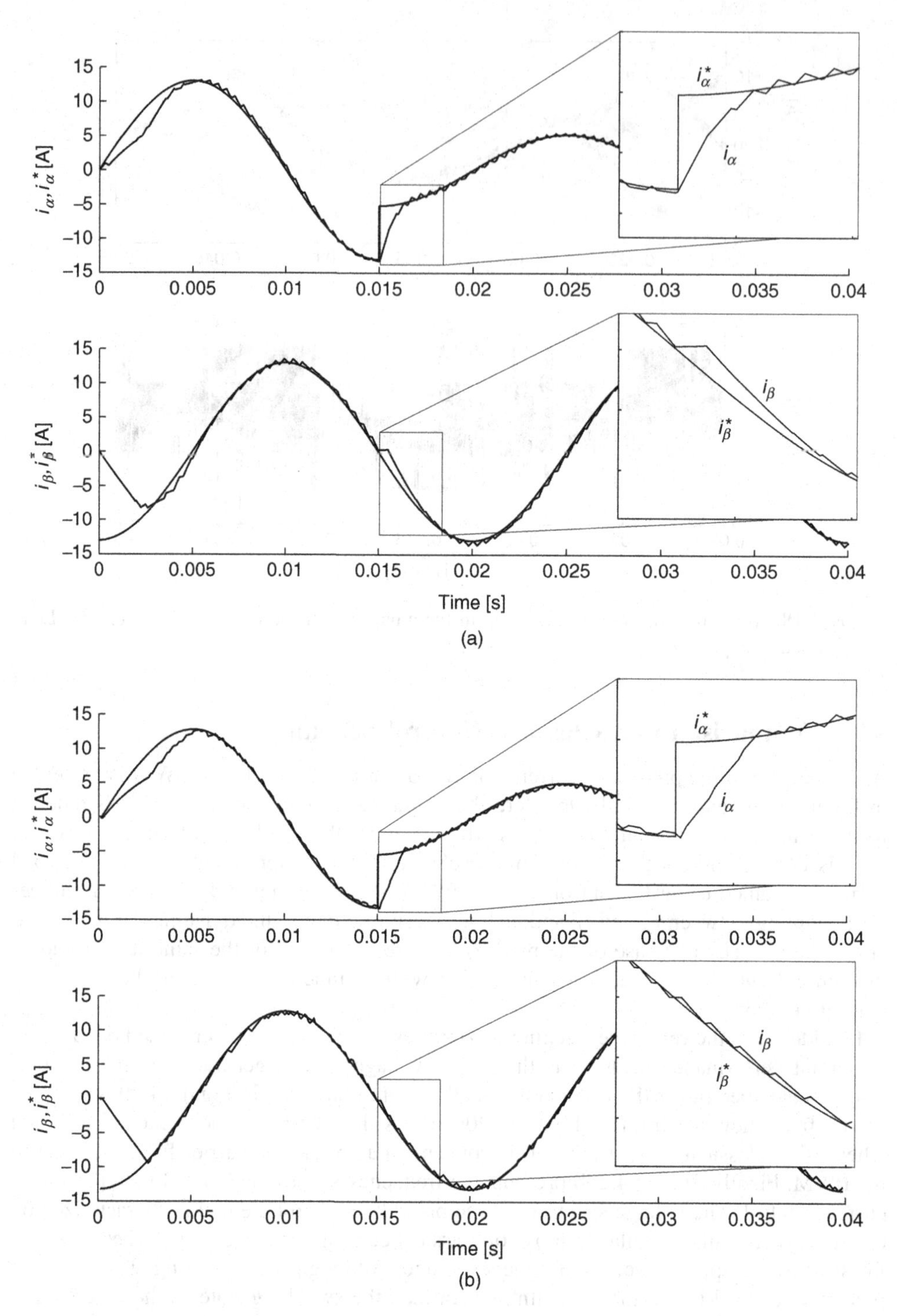

Figure 4.19 Results for a step in the reference current i_α^*. (a) Classical current control with PWM. (b) Predictive current control (Rodriguez *et al.*, 2007 © IEEE)

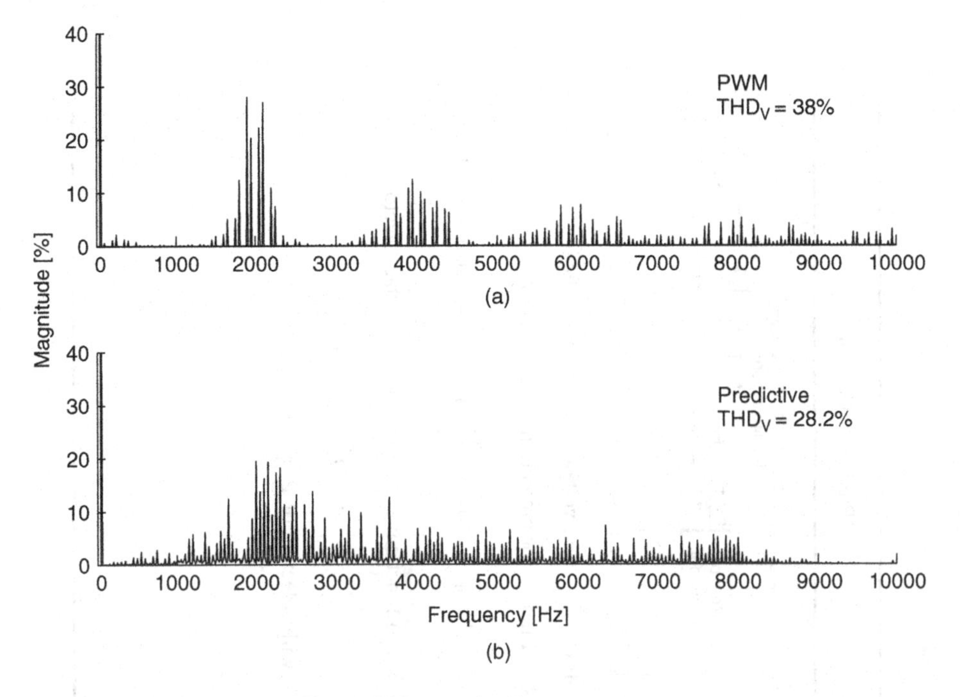

Figure 4.20 Load voltage spectrum

sampling instant, therefore the average switching frequency is always less than $f_s/2$. Results show that the average switching frequency is concentrated between $f_s/5$ and $f_s/4$. Additionally, the switching frequency can be controlled by modification of the cost function as will be shown in the next application example.

An assessment of the operating principles between a classical current control scheme and a predictive current control scheme is presented in Table 4.3. Although the predictive control scheme is based on more advanced control theory, the resulting control strategy is no more complex than a classical scheme based on PI controllers and SVM. Both control schemes need a model of the inverter and the voltage vectors that it generates. In the classical scheme, knowledge of the voltage vectors is used for implementation of the modulator. For predictive control, these voltage vectors are the finite set of possible actuations. In order to adjust the PI controllers, a linear model of the load is needed. The predictive controller will calculate predictions for each voltage vector using a discrete-time model of the load, which does not need to be linear. The performance of the PI controllers depends on the appropriate adjustment of their parameters k_p and k_i. In the predictive control scheme there are no parameters to adjust, but a cost function must be defined, which in the case of current control is very simple. The space vector modulator must be implemented by considering the calculation of the application times of the selected voltage vectors as explained in Section 2.1.2. This stage is not needed in the predictive control scheme as the switching states are generated directly by the controller.

Table 4.3 Assessment of operating principles

<table>
<tr><th>Current Control with SVM</th><th>Predictive Control</th></tr>
<tr><td>
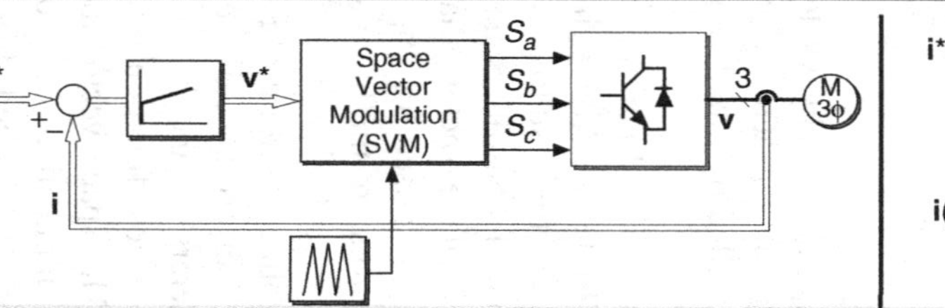

</td><td></td></tr>
<tr><td>1. SVM uses seven voltage vectors for modulation</td><td>1. MPC uses seven voltage vectors for prediction</td></tr>
<tr><td>2. Linear model
$$\mathbf{v} = R\mathbf{i} + L\frac{d\mathbf{i}}{dt} + \mathbf{e}$$</td><td>2. Load model for prediction (can be nonlinear):
$$\mathbf{i}^{\mathbf{p}}(k+1) = \left(1 - \frac{RT_s}{L}\right)\mathbf{i}(k) + \frac{T_s}{L}\left(\mathbf{v}(k) - \hat{\mathbf{e}}(k)\right)$$</td></tr>
<tr><td>3. Controller adjustment:
$$v^* = k_p(\mathbf{i}^* - \mathbf{i}) + k_i\int(\mathbf{i}^* - \mathbf{i})dt$$</td><td>3. Cost to function: evaluated for each voltage vector:
$$g = |i_\alpha^* - i_\alpha^{\mathbf{p}}| + |i_\beta^* - i_\beta^{\mathbf{p}}|$$
with selection of g optimal.</td></tr>
<tr><td>4. Space vector modulation (SVM)
$$\mathbf{v}^* = \frac{1}{T}(\mathbf{v}_a t_a + \mathbf{v}_b t_b + \mathbf{v}_0 t_0)$$
$$T = t_a + t_b + t_0$$</td><td></td></tr>
</table>

4.10 Summary

This chapter presents one of the simplest predictive control schemes: current control in a three-phase inverter. This example allows us to introduce this control method in the field of power electronics. A simple model of the converter and the load is presented. Then the predictive current control method is explained in detail. Results are presented and compared to a classical current control scheme.

References

[1] J. Holtz, "Pulsewidth modulation electronic power conversion," *Proceedings of the IEEE*, vol. 82, no. 8, pp. 1194–1214, August 1994.

[2] M. P. Kazmierkowski, R. Krishnan, and F. Blaabjerg, *Control in power electronics*. Academic Press, 2002.

[3] N. Mohan, T. M. Undeland, and W. P. Robbins, *Power electronics*, 3rd ed. John Wiley & Sons, Inc., 2003.

[4] J. Rodríguez, J. Pontt, C. Silva *et al.* "Predictive current control of a voltage source inverter," *IEEE Transactions on Industrial Electronics*, vol. 54, no. 1, pp. 495–503, February 2007.

4.10 Summary

[illegible]

References

[illegible]

5

Predictive Control of a Three-Phase Neutral-Point Clamped Inverter

5.1 Introduction

Three-level neutral-point clamped (NPC) inverters are widely used in industry for high-power, medium-voltage power conversion and drives [1, 2]. Topics related to power losses due to commutation and quality of the output current are relevant issues in this power range [3–5]. The neutral-point balancing problem in this topology is another subject that has been studied in recent years [6–8]. Among the most common control methods for this converter, the literature states, are non linear techniques, like hysteresis control, and linear methods, like the use of PI controllers in conjunction with pulse width modulation (PWM) [9–12].

The general predictive control scheme presented in Chapter 3 is applied here to the NPC inverter. The behavior of the system is predicted for each possible switching state of this kind of inverter. The switching state that minimizes a given cost function is selected to be applied during the next sampling interval following the same strategy presented in Chapters 3 and 4. The NPC inverter presents a high number of switching states, compared to the two-level inverter used in the previous chapter. The larger set of possible actuations allows for additional degrees of freedom and several compositions of the cost function can be considered. Considering the control requirements that are characteristic of this topology and its applications, several variations of the algorithm are studied and compared to classical linear control with PWM, including features like load current reference tracking, balance in the DC link capacitor voltages, and reduction of the switching frequency [13].

Predictive Control of Power Converters and Electrical Drives, First Edition. Jose Rodriguez and Patricio Cortes.

5.2 System Model

The power circuit of the NPC inverter is shown in Figure 5.1. Each phase of the inverter is composed of four switches and two diodes where the two center switches and the diodes allow the output terminal to be connected to the neutral point of the DC link. This configuration allows the generation of three voltage levels at the output terminal of phase x, with respect to the neutral point 0, considering the switching combinations given in Table 5.1.

Switching state variable S_x represents the switching state of phase x, with $x = \{a, b, c\}$, and it has three possible values denoted by $+$, 0, and $-$ that represent the switching combinations that generate $V_{dc}/2$, 0, and $-V_{dc}/2$, respectively, at the output of the inverter phase.

For the three phases of the inverter, 27 switching states are generated, which produce 19 different voltage vectors, as shown in Figure 5.2. Note that some switching states are

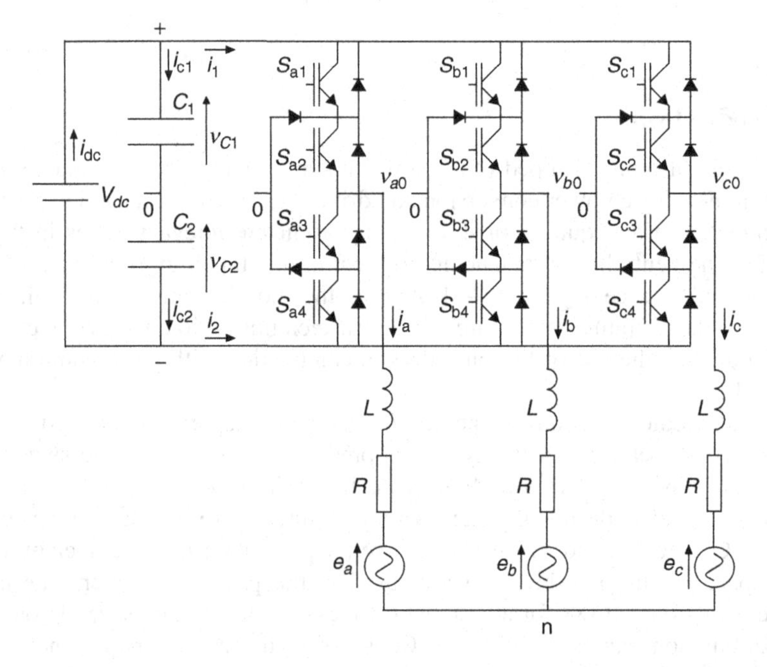

Figure 5.1 Circuit of a three-phase NPC inverter connected to a resistive–inductive–active load (Vargas *et al.*, 2007 © IEEE)

Table 5.1 Switching states for one phase of the inverter

S_x	S_{x1}	S_{x2}	S_{x3}	S_{x4}	v_{x0}
+	1	1	0	0	$V_{dc}/2$
0	0	1	1	0	0
−	0	0	1	1	$-V_{dc}/2$

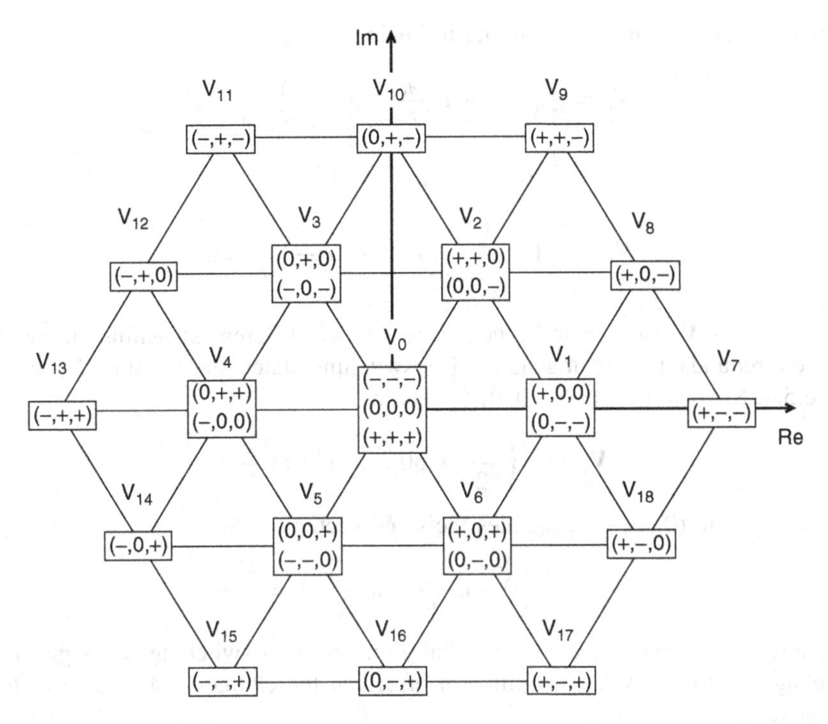

Figure 5.2 Possible voltage vectors and switching states generated by a three-level inverter (Vargas *et al.*, 2007 © IEEE)

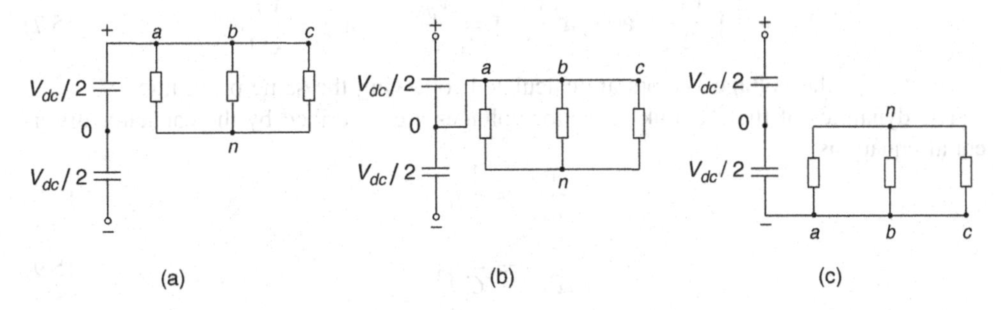

Figure 5.3 Different switching states that generate the zero vector $\mathbf{V}_0$. (a) (+, +, +). (b) (0, 0, 0). (c) (–, –, –)

redundant, generating the same voltage vector. For example, vector $\mathbf{V}_0$ can be generated by three different switching states: (+, +, +), (0, 0, 0), and (–, –, –), which generate the load configurations shown in Figure 5.3. Considering the space vector definition for the output voltage

$$\mathbf{v} = \frac{2}{3}\left(v_{a0} + \mathbf{a}v_{b0} + \mathbf{a}^2 v_{c0}\right) \tag{5.1}$$

these three switching states generate the following voltage vectors:

$$\mathbf{V}_0 = \frac{2}{3}\left(\frac{V_{dc}}{2} + \mathbf{a}\frac{V_{dc}}{2} + \mathbf{a}^2\frac{V_{dc}}{2}\right) = 0 \tag{5.2}$$

$$\mathbf{V}_0 = \frac{2}{3}\left(0 + \mathbf{a}0 + \mathbf{a}^2 0\right) = 0 \tag{5.3}$$

$$\mathbf{V}_0 = \frac{2}{3}\left(-\frac{V_{dc}}{2} - \mathbf{a}\frac{V_{dc}}{2} - \mathbf{a}^2\frac{V_{dc}}{2}\right) = 0 \tag{5.4}$$

Voltage vectors $\mathbf{V}_1$ to $\mathbf{V}_6$ can be generated by two different switching states, that is, they present redundant switching states. The switching states that generate $\mathbf{V}_1$ are shown in Figure 5.4. Switching state (+, 0, 0) results in

$$\mathbf{V}_1 = \frac{2}{3}\left(\frac{V_{dc}}{2} + \mathbf{a}0 + \mathbf{a}^2 0\right) = \frac{V_{dc}}{3} \tag{5.5}$$

and switching state (0, −, −) generates the same vector

$$\mathbf{V}_1 = \frac{2}{3}\left(0 - \mathbf{a}\frac{V_{dc}}{2} - \mathbf{a}^2\frac{V_{dc}}{2}\right) = \frac{V_{dc}}{3} \tag{5.6}$$

It can also be observed in Figure 5.4 that although both switching states generate the same voltage vector, they have a different effect on the charge or discharge of the DC link capacitors.

Outer vectors present no redundancies. Figure 5.5 shows switching state (+, 0, −) that generates vector $\mathbf{V}_8$, calculated as

$$\mathbf{V}_8 = \frac{2}{3}\left(\frac{V_{dc}}{2} + \mathbf{a}0 - \mathbf{a}^2\frac{V_{dc}}{2}\right) = \frac{V_{dc}}{3}(1 - \mathbf{a}^2) = \frac{V_{dc}}{\sqrt{3}}e^{j\pi/6} \tag{5.7}$$

The rest of the voltage vectors are calculated following the same procedure.

The dynamics of the DC link capacitor voltages are described by the capacitor differential equations

$$\frac{dv_{c1}}{dt} = \frac{1}{C}i_{c1} \tag{5.8}$$

$$\frac{dv_{c2}}{dt} = \frac{1}{C}i_{c2} \tag{5.9}$$

where C is the capacitor value.

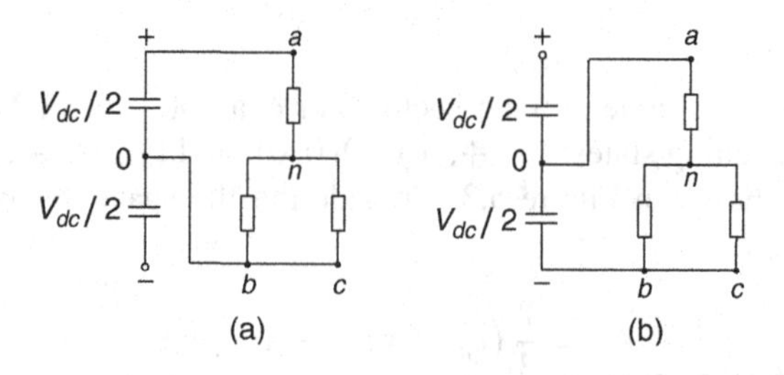

Figure 5.4 Different switching states that generate vector $\mathbf{V}_1$. (a) (+, 0, 0). (b) (0, −, −)

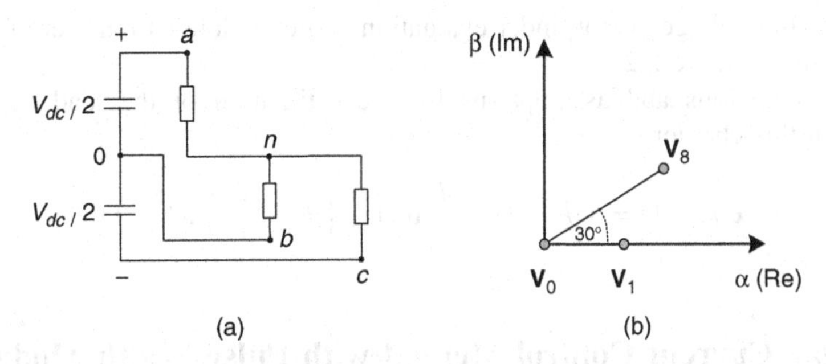

Figure 5.5 Generation of voltage vector vector $\mathbf{V}_8$. (a) Switching configuration (+, 0, −). (b) Vector $\mathbf{V}_8$ in the complex plane

The same approximation of the derivative considered in the previous chapter can be used for the capacitor voltages for a sampling time T_s,

$$\frac{dv_{cx}}{dt} \approx \frac{v_{cx}(k+1) - v_{cx}(k)}{T_s} \tag{5.10}$$

giving the following discrete-time equations:

$$v^{\mathbf{p}}_{c1}(k+1) = v_{c1}(k) + \frac{1}{C} i_{c1}(k) T_s \tag{5.11}$$

$$v^{\mathbf{p}}_{c2}(k+1) = v_{c2}(k) + \frac{1}{C} i_{c2}(k) T_s \tag{5.12}$$

where currents $i_{c1}(k)$ and $i_{c2}(k)$ depend on the switching state of the inverter and the value of the output currents, and can be calculated using the following expressions:

$$i_{c1}(k) = i_{dc}(k) - H_{1a} i_a(k) - H_{1b} i_b(k) - H_{1c} i_c(k) \tag{5.13}$$

$$i_{c2}(k) = i_{dc}(k) + H_{2a} i_a(k) + H_{2b} i_b(k) + H_{2c} i_c(k) \tag{5.14}$$

where i_{dc} is the current supplied by the voltage source V_{dc}. Variables H_{1x} and H_{2x} depend on the switching states and are defined as

$$H_{1x} = \begin{cases} 1 & \text{if } S_x = \text{“}+\text{”} \\ 0 & \text{otherwise} \end{cases} \tag{5.15}$$

$$H_{2x} = \begin{cases} 1 & \text{if } S_x = \text{“}-\text{”} \\ 0 & \text{otherwise} \end{cases} \tag{5.16}$$

with $x = a, b, c$.

Hence, (5.11)–(5.14) allow us to predict the effect of selecting a given switching state on the variation of the capacitor voltages.

The same three-phase resistive–inductive–active load considered in the previous chapter is used with the NPC inverter. The discrete-time model for the load current vector is expressed as

$$\mathbf{i}^{\mathbf{p}}(k+1) = \left(1 - \frac{RT_s}{L}\right) \mathbf{i}(k) + \frac{T_s}{L} \left(\mathbf{v}(k) - \hat{\mathbf{e}}(k)\right) \tag{5.17}$$

where $\mathbf{v}(k)$ is the voltage vector under evaluation, which belongs to the set of 19 voltage vectors shown in Figure 5.2.

The same equations and assumptions for the estimation of the load back-emf are considered in this chapter

$$\hat{\mathbf{e}}(k-1) = \mathbf{v}(k-1) - \frac{L}{T_s}\mathbf{i}(k) - \left(R - \frac{L}{T_s}\right)\mathbf{i}(k-1) \tag{5.18}$$

5.3 Linear Current Control Method with Pulse Width Modulation

Before presenting the predictive control method, a short review of classical current control applied to a three-phase NPC inverter is given to obtain suitable comparisons. The selected method involves linear controllers and a modulation strategy known as level shifted phase disposition pulse width modulation (LS-PD-PWM). This alternative was selected, among other PWM strategies, because it is widely used on this kind of inverter and provides the best harmonic profile [1].

The classical control scheme using PWM is shown in Figure 5.6. The load current is measured and compared to its reference value. Next, a PI controller generates the reference load voltages that enter a modulator. In this stage, each reference voltage is compared to two triangular carrier signals, superior and inferior, arranged in identical phase disposition. The switching state applied to the inverter is selected according to the results of the comparisons. For more details, see [2, 9–11].

5.4 Predictive Current Control Method

The predictive control scheme for the NPC inverter is shown in Figure 5.7. The future values of the load currents and voltages in the capacitors are predicted for the 27 switching states generated by the inverter, by means of (5.17), (5.11), and (5.12). For this purpose, it is necessary to measure the present load currents and voltages in the capacitors. After obtaining the predictions, a cost function g is evaluated for each switching state. The switching state that minimizes the cost function is selected and applied during the next sampling period.

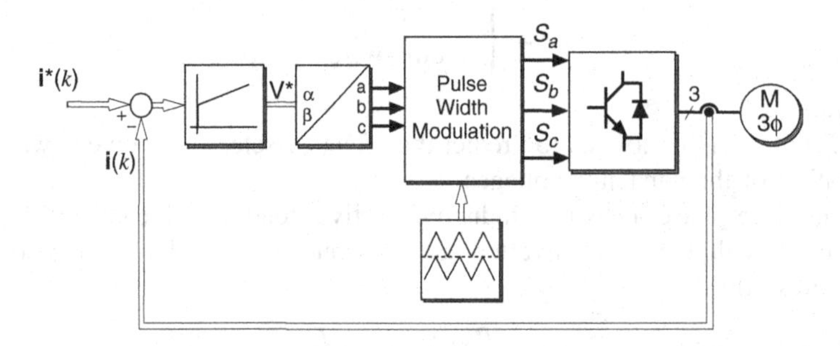

Figure 5.6 Classical current control method for the NPC inverter (Vargas *et al.*, 2007 © IEEE)

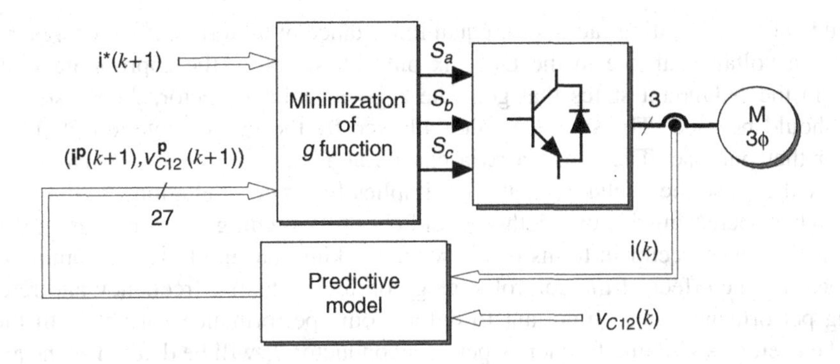

Figure 5.7 Predictive current control method for the NPC inverter (Vargas *et al.*, 2007 © IEEE)

The control requirements for the NPC inverter are:

- Load current reference tracking
- DC link capacitor voltages balance
- Reduction of the switching frequency.

These requirements can be formulated in the form of a cost function to be minimized. The cost function for the NPC inverter has the following composition:

$$g = |i_\alpha^* - i_\alpha^{\mathbf{p}}| + |i_\beta^* - i_\beta^{\mathbf{p}}| + \lambda_{dc}|v_{c1}^{\mathbf{p}} - v_{c2}^{\mathbf{p}}| + \lambda_n n_c \tag{5.19}$$

The first two terms are the load current errors in orthogonal coordinates, where $i_\alpha^{\mathbf{p}}$ and $i_\beta^{\mathbf{p}}$ are the real and imaginary components of the predicted current vector $\mathbf{i}^{\mathbf{p}}$, respectively, and i_α^* and i_β^* are the real and imaginary components of the reference current vector $\mathbf{i}^*$, as defined in the previous chapter.

The third term in the cost function measures the difference in the predicted values of the DC link capacitor voltages. These predicted voltages are calculated using (5.11) and (5.12). Then, by minimization of this term, the capacitor voltages will tend to be equal.

The last term is the number of commutations required to switch from the present switching state to the switching state under evaluation. A switching state that implies fewer commutations of the power semiconductors will be preferred. In this manner, the use of this term will have a direct effect on the switching frequency of the converter.

Weighting factors λ_{dc} and λ_n handle the relation between terms dedicated to reference tracking, voltage balance, and reduction of switching frequency within the cost function g. A large value of certain λ implies greater priority to that objective.

The basic predictive current control strategy, applying cost function (5.19) with $\lambda_{dc} = \lambda_n = 0$, requires no parameter adjustment, only knowledge of the load. Nevertheless, to take advantage of the possibilities offered by this control method, it is necessary to adjust parameters λ_{dc} and λ_n. No design procedure has been established thus far for this purpose. However, some general guidelines on weighting factor selection can be found in Chapter 11. First, the designer should consider the different units and magnitudes of the variables involved in the cost function g. This will give some idea about the order of magnitude of the weighting factors for equal importance of all terms. If the designer wants to maintain voltage balance in the DC link only by selecting the appropriate switching state from the redundant states that generate a given voltage vector, then a small value of λ_{dc} should be used. The smallest value allowed by the implementation platform will work for that purpose. The same criteria can be applied to λ_n. With a small value, the method will choose the switching state that implies fewer commutations within a voltage vector. When increasing λ_n, the method could choose switching states that are not within the optimal voltage vector in terms of reference tracking, but imply fewer commutations.

To measure the effect of the control strategy on the switching frequency and reference tracking performance, it is important to define some performance variables. In the first place, the average switching frequency per semiconductor f_s will be defined as the average value of the switching frequencies of the 12 controlled power semiconductors in the converter circuit. Accordingly,

$$f_s = \sum_{i=1}^{4} \frac{f_{s_{ai}} + f_{s_{bi}} + f_{s_{ci}}}{12} \tag{5.20}$$

where $f_{s_{ki}}$ is the average switching frequency, during a time interval, of power semiconductor number i of phase k, with $i \in \{1, 2, 3, 4\}$ and $k \in \{a, b, c\}$. A reduction in the switching frequency of the inverter will imply a reduction in f_s. As the reader can observe, f_s was defined as an average between switching frequencies. Not all 12 power semiconductors will present the same switching frequency. Moreover, transitions will occur in general with different current values, so f_s will not be directly proportional to the power losses in the converter. However, it will allow us to measure or get some indication of the switching frequency of the inverter and the power losses due to commutations.

The mean absolute reference tracking error $\overline{e}$ will be defined as the mean value of the absolute differences between the reference current and the measured load current, within a given time interval of m samples:

$$\overline{e} = \frac{1}{m} \sum_{k=0}^{m} |i^*(k) - i(k)| \tag{5.21}$$

As a difference between currents it will be measured in amperes and will also be expressed as a percentage of the amplitude of the reference current.

5.5 Implementation

The control strategy was implemented based on a dSPACE DS1104 rapid prototyping system and MATLAB® and Simulink® installed on a host PC. The sampling period used with the predictive strategy was $T_s = 100\,\mu s$, or a 10 kHz sampling frequency. The predictive algorithm implemented with the control platform based on the dSPACE DS1104 is explained in a flow diagram presented in Figure 5.8. The control loop begins sampling the required signals. Then, the algorithm estimates the active component of the load by means of (5.18) and initializes the value of g_{op}, a variable that will contain the value of the lower cost function evaluated by the algorithm so far. Then the strategy enters a loop where, for each possible switching state, the cost function (5.19) is evaluated considering

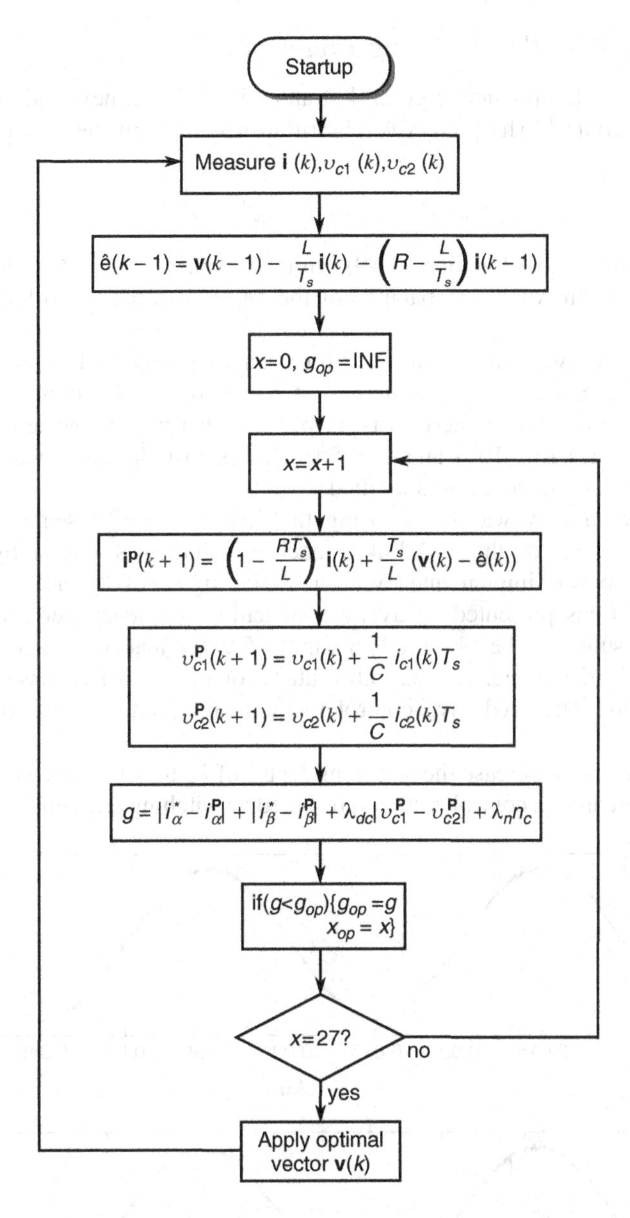

Figure 5.8 Flow diagram of the implemented control algorithm

current and voltage predictions obtained from (5.17), (5.11), and (5.12), respectively. If, for a given switching state, the evaluated cost function g happens to be lower than g_{op}, that lower value is stored as g_{op} and the switching state number is stored as j_{op}. The loop ends when all 27 switching states have been evaluated. The state that produces the optimal value of g (minimal) is identified by variable j_{op} and will be applied to the converter during the next sampling interval, starting the control algorithm again.

5.5.1 Reduction of the Switching Frequency

The performance of the predictive control strategy is analyzed next and compared to the classical current control. The predictive algorithm was implemented using the following cost function:

$$g = |i_\alpha^* - i_\alpha^p| + |i_\beta^* - i_\beta^p| + \lambda_n n_c \tag{5.22}$$

which corresponds to cost function (5.19) with $\lambda_{dc} = 0$. This cost function can be considered for systems in which the balance of the DC link capacitor voltages is provided by the rectifier.

The PWM method was implemented with carrier frequencies of 1440 Hz and 400 Hz. The total DC link voltage was maintained at 533 V by a DC source that also maintained voltage balance during these tests. A passive load was connected to the inverter, with parameter values $R = 10\,\Omega$ and $L = 50$ mH. A sinusoidal reference current of 10 A amplitude and 50 Hz frequency was applied.

The predictive strategy was tested using the cost function presented in (5.22) with $\lambda_n = 0.001$. Balancing of the DC link capacitor voltages is forced by the rectifier. The PWM method was implemented with a carrier signal of frequency $f_c = 1440$ Hz. Both implementations presented an average switching frequency per semiconductor of $f_s = 720$ Hz. Results can be observed in Figure 5.9 for load current on phase a, and Figure 5.10 for load voltage. A mean absolute error of $\overline{e} = 0.184$ A was measured for the PWM strategy. The predictive control method presented a mean absolute error of $\overline{e} = 0.165$ A.

The second step is to increase the weighing factor of λ_n to 0.16, to reduce the switching frequency. The predictive control method presented a switching frequency of $f_s = 200$ Hz.

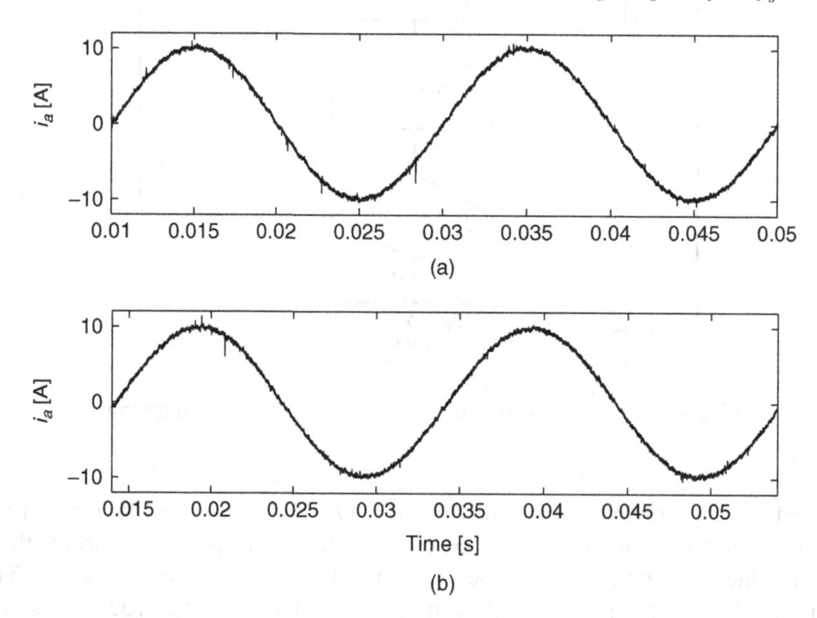

Figure 5.9 Results with $f_s = 720$ Hz, load current i_a A. (a) PWM. (b) Predictive (Vargas *et al.*, 2007 © IEEE)

The PWM method was adjusted to match the switching frequency, with a carrier signal of frequency $f_c = 400\,\text{Hz}$. Results on the load current for both methods can be observed in Figure 5.11. The load voltage signals for the PWM and predictive methods can be observed in Figure 5.12. Comparing Figures 5.11 and 5.12 to Figures 5.9 and 5.10, it is possible to verify the reduction in the switching frequency, as well as an increase in the reference tracking error for both methods. Analysis of the mean absolute error, however, reveals a significant difference in the performance of both methods. The PWM strategy presented a mean absolute error of $\overline{e} = 0.406\,\text{A}$, while the predictive method achieved a value of $\overline{e} = 0.283\,\text{A}$, both working at $f_s = 200\,\text{Hz}$.

Table 5.2 presents a review summarizing the most relevant characteristics and results for both methods, including average switching frequency per IGBT f_s, mean absolute tracking error $\overline{e}$, and sampling frequency required to apply the method. The theoretical maximum switching frequency that each method can reach will depend basically on the sampling frequency. For the PWM method, the theoretical maximum f_s is equal to the sampling frequency used, whereas for the predictive strategy the theoretical maximum f_s is equal to half the sampling frequency. These values limit the switching frequency.

To reveal the possibilities of the predictive control method, a graph showing the relation between the design parameter λ_n and the average switching frequency per semiconductor f_s and mean absolute reference tracking error $\overline{e}$ is presented in Figure 5.13. From that figure, built on several simulations for each value of λ_n, it is possible to confirm the relation mentioned. Increasing λ_n implies a reduction in the switching frequency and increases the reference tracking error. The designer should select λ_n and λ_{dc} to fit the requirements in terms of switching frequency and reference tracking.

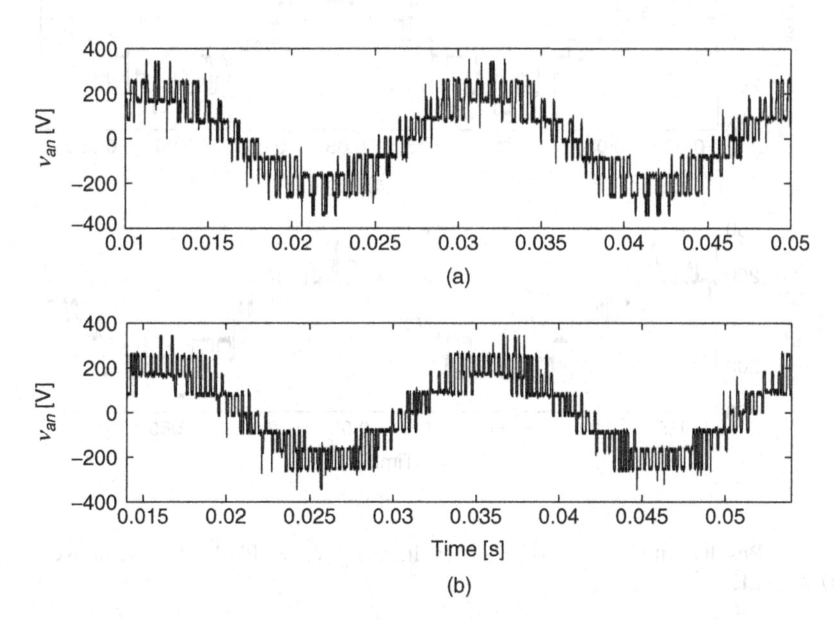

Figure 5.10 Results with $f_s = 720\,\text{Hz}$, load voltage v_{an} V. (a) PWM. (b) Predictive control (Vargas *et al.*, 2007 © IEEE)

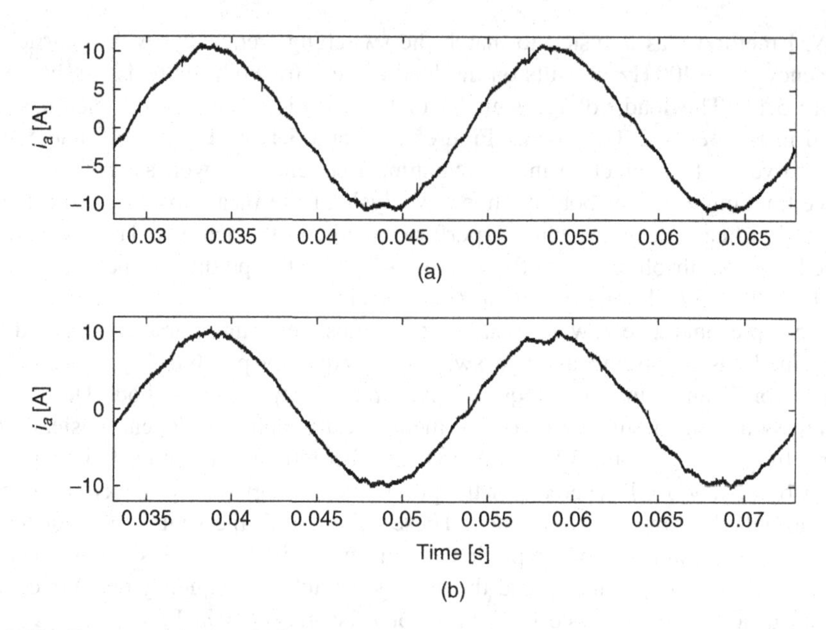

Figure 5.11 Results with $f_s = 200\,\text{Hz}$, load current i_a A. (a) PWM. (b) Predictive (Vargas *et al.*, 2007 © IEEE)

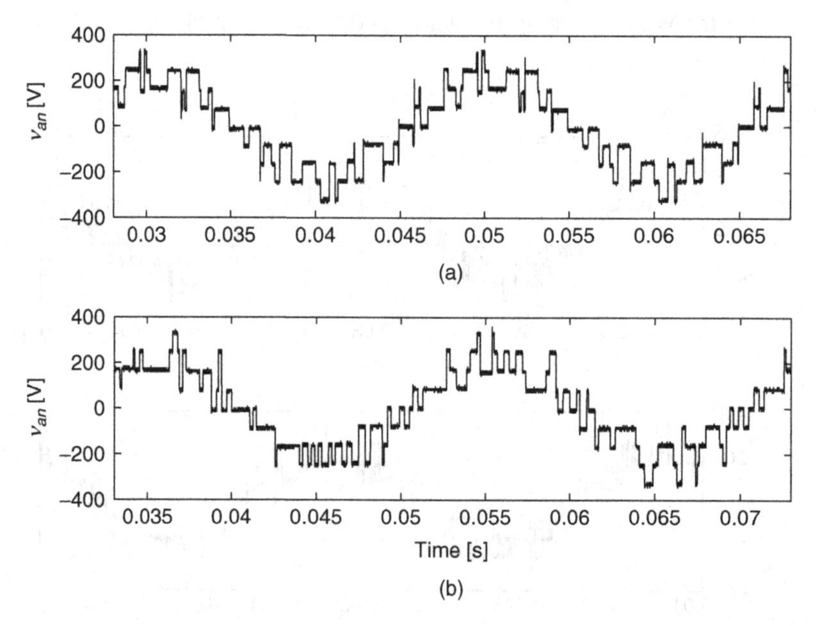

Figure 5.12 Results with $f_s = 200\,\text{Hz}$, load voltage v_{an} V. (a) PWM. (b) Predictive (Vargas *et al.*, 2007 © IEEE)

Table 5.2 Comparative performance of PWM and predictive methods (Vargas *et al.*, 2007 © IEEE)

Control method	Switching frequency [Hz]	Absolute error [A]	Sampling frequency [kHz]
PWM $f_c = 1440$ [hz]	720	0.184	1.44
Predictive $\lambda_n = 0.001$	720	0.165	10
PWM $f_c = 400$ [hz]	200	0.406	0.4
Predictive $\lambda_n = 0.16$	200	0.283	10

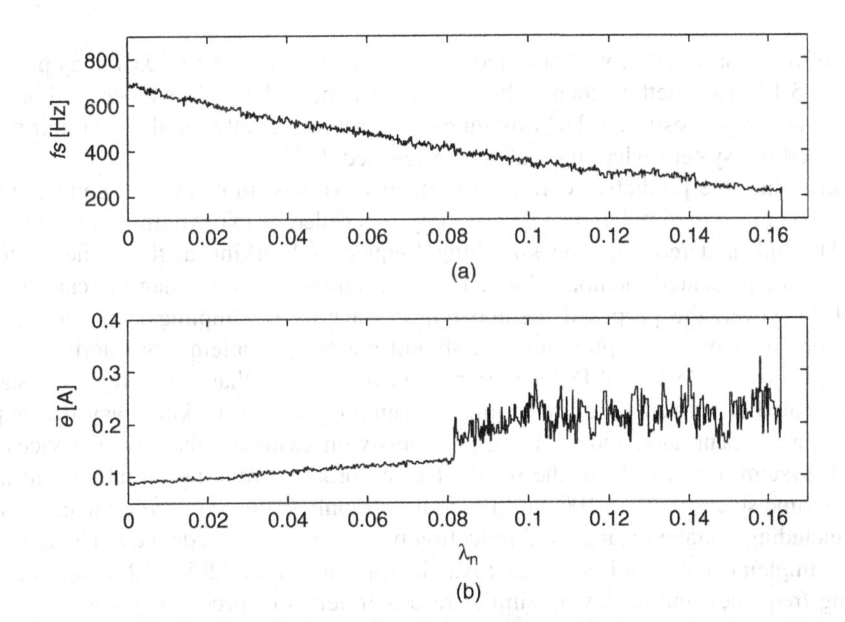

Figure 5.13 Design parameter λ_n. (a) Relation with the switching frequency [Hz] (switching frequency vs. λ_n). (b) Relation with the absolute error [A] (absolute error vs. λ_n) (Vargas *et al.*, 2007 © IEEE)

5.5.2 Capacitor Voltage Balance

One of the most interesting aspects of the predictive control method is the simplicity for implementing voltage balance in the DC link. This feature was tested by disconnecting the middle point of the DC link from the source and applying the predictive control method with the following cost function:

$$g = |i_\alpha^* - i_\alpha^{\mathbf{p}}| + |i_\beta^* - i_\beta^{\mathbf{p}}| + \lambda_{dc} \left| v_{c1}^{\mathbf{p}} - v_{c2}^{\mathbf{p}} \right| \tag{5.23}$$

The λ_{dc} weighing factor was set at $\lambda_{dc} = 0.1$. The method succeeded in maintaining voltage balance, using the same reference signal and parameters as in the previous experimental implementations. To show the capabilities of the method, the voltage balance

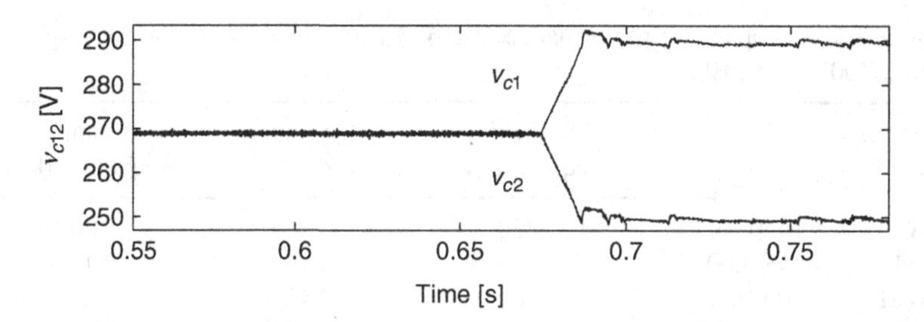

Figure 5.14 Test regarding voltage balance in the DC link capacitors applying the predictive strategy (Vargas *et al.*, 2007 © IEEE)

section of the cost function was disabled, setting $\lambda_{dc} = 0$ at time $t = 0.67$ s, as presented in Figure 5.14. The method then will not consider the voltage unbalance within g. As expected, both voltages in the DC link quickly began to separate, until the circuit protection stopped the system when the unbalance reached 40 V.

Summarizing, the predictive current control method was implemented, confirming the observations made in simulations. The strategy succeeded in maintaining voltage balance in the DC link and reducing the switching frequency. Working at the same switching frequency, the presented method achieved better reference tracking than the carrier-based method. However, the proposed method requires a greater sampling frequency or data acquisition frequency. The previous fact should not be a problem, considering the new technologies available for DSPs. It is important to mention that the sampling instant is always located at a fixed position within the sampling period, making easy the acquisition of measurement data, and avoiding problems with switching the power devices. The dSPACE system used to obtain the results had no problem running the algorithm at the sampling time selected, $T_s = 100\,\mu$s. In fact, it took only 52 μs to execute the entire algorithm, including voltage balance and reduction of the switching frequency. The algorithm was also implemented on a DSP from Texas Instruments, TMS320F2812, using the same sampling frequency and achieving similar results in terms of processing times.

One of the aspects that must be mentioned is the simplicity of implementing the voltage balance strategy with the presented method. There is no need to consider long look-up tables or additional control blocks.

5.6 Summary

The predictive current control method presented does not require any kind of linear controller or modulation technique. It effectively controls the load current and compares well to established control methods like PWM, achieves a comparable dynamic response and reference tracking, and works at lower switching frequencies. If both methods are compared at the same switching frequency, as presented in Table 5.2, the predictive strategy reveals a lower tracking error. In addition, the proposed method shows no interaction between both components of the load current.

One of the remarkable aspects of the method is the use of costs assigned to each objective to achieve reference tracking, balance in the DC link, and a reduction in the switching frequency. The simplicity of the theory makes it easy to understand and implement. The strategy allows the designer to adjust the λ parameters to fit the requirements in terms of switching frequency, voltage balance, and reference tracking. A systematic way to determine the weighting factors is a challenge for future work. Some guidelines on the adjustment of these parameters are given in Chapter 11.

The method can be easily implemented by taking advantage of the present technologies available for DSPs. The higher sampling frequencies required should not be a problem nowadays. This control strategy uses, in a very convenient way, the discrete nature of power converters and microprocessors used in their control.

References

[1] A. Nabae, I. Takahashi, and H. Akagi, "A new neutral-point-clamped PWM inverter," *IEEE Transactions on Industry Applications*, vol. IA-17, no. 5, pp. 518–523, September/October 1981.

[2] B. Wu, "High-power converters and AC motor drives," Power Electronics Specialists Conference, PESC'05, 2005.

[3] H.-P. Krug, T. Kume, and M. Swamy, "Neutral-point clamped three-level general purpose inverter – features, benefits and applications," Power Electronics Specialists Conference, PESC'04, 2004.

[4] M. K. Buschmann and J. K. Steinke, "Robust and reliable medium voltage PWM inverter with motor friendly output," 7th European Conference on Power Electronics and Applications, pp. 3502–3507, 1997.

[5] T. Bruckner, S. Bernet, and H. Guldner, "The active NPC converter and its loss-balancing control," *IEEE Transactions on Industrial Electronics*, vol. 52, no. 3, pp. 855–868, June 2005.

[6] R. Tallam, R. Naik, and T. Nondahl, "A carrier-based PWM scheme for neutral-point voltage balancing in three-level inverters," *IEEE Transactions on Industry Applications*, vol. 41, no. 6, pp. 1734–1743, November/December 2005.

[7] A. Bendre, G. Venkataramanan, D. Rosene, and V. Srinivasan, "Modeling and design of a neutral-point voltage regulator for a three-level diode-clamped inverter using multiple-carrier modulation," *IEEE Transactions on Industrial Electronics*, vol. 53, no. 3, pp. 718–726, June 2006.

[8] H. du Toit Mouton, "Natural balancing of three-level neutral-point-clamped PWM inverters," *IEEE Transactions on Industrial Electronics*, vol. 49, no. 5, pp. 1017–1025, October 2002.

[9] G. Holmes and T. Lipo, Pulse width modulation for power converters: Principles and practice, IEEE Press Series on Power Engineering, Wiley-Interscience, 2003.

[10] M. Kazmierkowski, R. Krishnan, and F. Blaabjerg, *Control in power electronics*, Academic Press, 2002.

[11] J. Holtz, "Pulsewidth modulation for electronic power conversion," *Proceedings of the IEEE*, vol. 82, no. 8, pp. 1194–1214, August 1994.

[12] J. E. Espinoza, J. R. Espinoza, and L. A. Moran, "A systematic controller-design approach for neutral-point-clamped three-level inverters," *IEEE Transactions on Industrial Electronics*, vol. 52, no. 6, pp. 1589–1599, December 2005.

[13] R. Vargas, P. Cortés, U. Ammann, J. Rodríguez, and J. Pontt, "Predictive control of a three-phase neutral-point-clamped inverter," *IEEE Transactions on Industrial Electronics*, vol. 54, no. 5, pp. 2697–2705, October 2007.

6

Control of an Active Front-End Rectifier

6.1 Introduction

Rectifiers are by far the most widely used converters in power electronics. The transformation from alternating current to direct current performed by rectifiers is used in a large variety of applications and from small power up to several megawatts.

The diode rectifier shown in Figure 6.1a is the simplest topology, which produces a fixed DC voltage, while the diodes are commutated by the AC voltages. This circuit is also known as a line-commutated rectifier and the power semiconductors operate at very low switching frequency. The main advantages of the diode rectifier are its simplicity and extremely low cost. The disadvantages and limitations of the three-phase diode rectifier are:

1. It does not offer the possibility of control of the power flow.
2. It generates high harmonics at the input current, especially when it supplies a capacitive load, as shown in Figure 6.1a. This capacitor is usually used to filter the output voltage.
3. It does not allow regeneration of power.

The second topology of importance is the thyristor rectifier, presented in Figure 6.1b, which introduces the possibility of control of power flow by changing the angle of the gate pulses (α) for the thyristors. Through this angle α it is possible to change the mean value of the load voltage, originating the control of power delivered to the load [1]. Thyristor rectifiers have in general the same advantages and limitations as diode rectifiers. An additional negative feature is that an increase in the value of α increases the phase displacement of the input current with respect to the source AC voltage, increasing the amount of fundamental reactive power. An advantage of thyristor rectifiers is that, in operating with $\alpha > 90°$, they can regenerate power from the DC load to the power supply.

The third important rectifier topology is presented in Figure 6.1c, which includes power transistors with antiparallel diodes as the main power switches. This rectifier operates

Predictive Control of Power Converters and Electrical Drives, First Edition. Jose Rodriguez and Patricio Cortes.

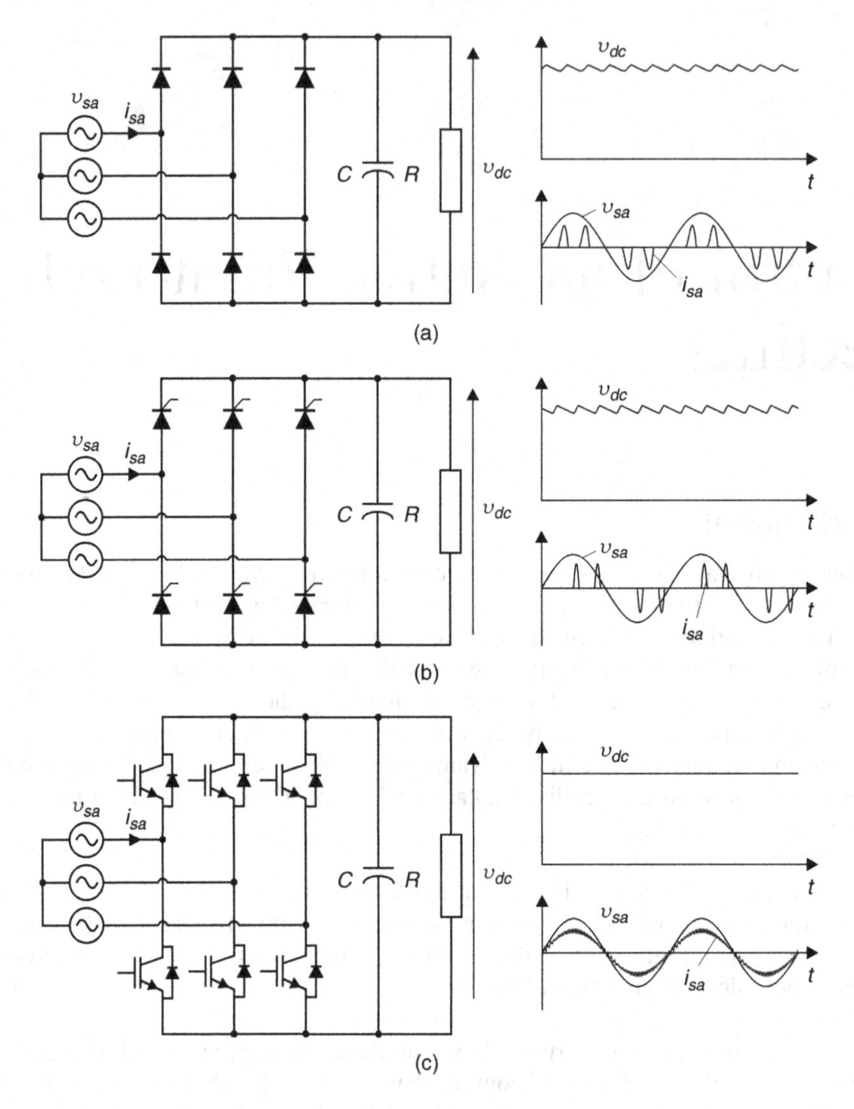

Figure 6.1 Three-phase rectifier. (a) Diode rectifier. (b) Thyristor rectifier. (c) AFE rectifier

with high switching frequency and is also known as an active front-end (AFE) rectifier. It overcomes all the problems and limitations of diode and thyristor rectifiers [2]. Its main features are:

1. Controlled DC voltage.
2. Controlled input currents with sinusoidal waveform (reduced harmonics).
3. Operation with very high power factor.
4. Full regenerative operation.

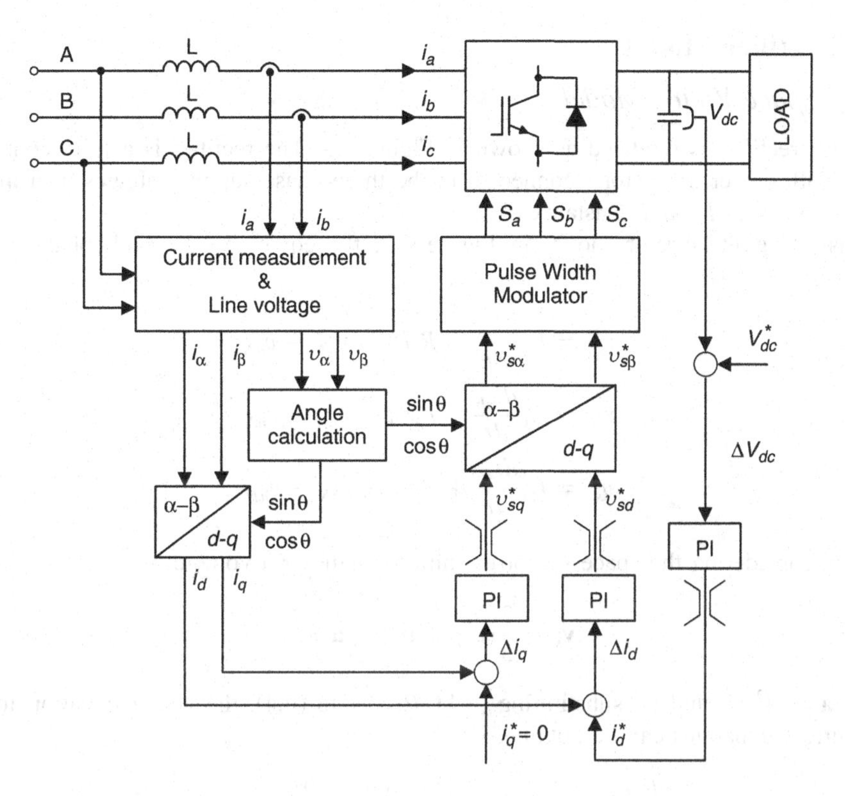

Figure 6.2 Voltage-oriented control (VOC) of an AFE rectifier

The most important disadvantage of this topology is the higher cost, in comparison to diode or thyristor rectifiers.

AFE rectifier control methods can be classified, as presented in [3], as voltage-oriented control (VOC) and direct power control (DPC). In VOC, input active power and reactive power are regulated indirectly by controlling the input currents, which are oriented with respect to the line voltage vector [4]. A VOC scheme considering current control in rotating coordinates is depicted in Figure 6.2. In this scheme, the angle of the grid voltage vectors is calculated for orientation of the rotating dq reference frame. In this way, the d-axis current i_d is proportional to the active power and the q-axis current i_q is proportional to the reactive power. In order to obtain unity power factor, current reference i_q^* is set to zero. The DC link voltage is regulated by a PI controller which generates the reference for the current related to the active power i_d^*. This method achieves good dynamic and static behavior; however, its performance depends on the quality of the current control strategies.

With DPC, active power and reactive power are estimated, using current measurements, and controlled directly with hysteresis controllers and a switching table similar to the one used in direct torque control (DTC) [5, 6].

In this chapter, the application of predictive control is presented for the control of an AFE considering the ideas of both the VOC and DPC schemes. The application of predictive power control for a regenerative drive is also presented in the chapter.

6.2 Rectifier Model

6.2.1 Space Vector Model

The AFE rectifier is modeled as shown in Figure 6.3. The rectifier is a fully controlled bridge with power transistors, connected to the three-phase supply voltages v_s using the filter inductances L_s and resistances R_s.

Considering the circuit shown in Figure 6.3, the equations for each phase can be written as

$$v_{sa} = L_s \frac{di_{sa}}{dt} + R_s i_{sa} + v_{aN} - v_{nN} \tag{6.1}$$

$$v_{sb} = L_s \frac{di_{sb}}{dt} + R_s i_{sb} + v_{bN} - v_{nN} \tag{6.2}$$

$$v_{sc} = L_s \frac{di_{sc}}{dt} + R_s i_{sc} + v_{cN} - v_{nN} \tag{6.3}$$

Then, considering the space vector definition for the grid voltage

$$\mathbf{v}_s = \frac{2}{3}(v_{sa} + \mathbf{a}v_{sb} + \mathbf{a}^2 v_{sc}) \tag{6.4}$$

where $\mathbf{a} = e^{j2\pi/3}$, and by substituting (6.1)–(6.3) into (6.4), the vector equation for the grid current dynamics can be obtained as

$$\begin{aligned}\mathbf{v}_s = {} & L_s \frac{d}{dt}\left(\frac{2}{3}(i_{sa} + \mathbf{a}i_{sb} + \mathbf{a}^2 i_{sc})\right) + R_s \frac{2}{3}(i_{sa} + \mathbf{a}i_{sb} + \mathbf{a}^2 i_{sc}) \\ & + \frac{2}{3}(v_{aN} + \mathbf{a}v_{bN} + \mathbf{a}^2 v_{cN}) - \frac{2}{3}(v_{nN} + \mathbf{a}v_{nN} + \mathbf{a}^2 v_{nN})\end{aligned} \tag{6.5}$$

Note that the last term of this equation is equal to zero

$$\frac{2}{3}(v_{nN} + \mathbf{a}v_{nN} + \mathbf{a}^2 v_{nN}) = v_{nN}\frac{2}{3}(1 + \mathbf{a} + \mathbf{a}^2) = 0 \tag{6.6}$$

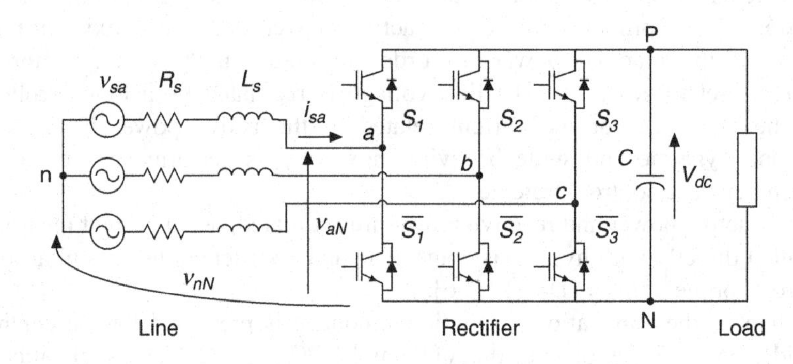

Figure 6.3 AFE rectifier

The input current dynamics equation (6.5) can be simplified by considering the following definitions for the grid current vector and the voltage vector generated by the AFE:

$$\mathbf{i}_s = \frac{2}{3}(i_{sa} + \mathbf{a}i_{sb} + \mathbf{a}^2 i_{sc}) \tag{6.7}$$

$$\mathbf{v}_{\text{afe}} = \frac{2}{3}(v_{aN} + \mathbf{a}v_{bN} + \mathbf{a}^2 v_{cN}) \tag{6.8}$$

Voltage $\mathbf{v}_{\text{afe}}$ is determined by the switching state of the converter and the DC link voltage, and can be expressed by the equation

$$\mathbf{v}_{\text{afe}} = \mathbf{S}_{\text{afe}} V_{dc} \tag{6.9}$$

where V_{dc} is the DC link voltage and $\mathbf{S}_{\text{afe}}$ is the switching state vector of the rectifier, defined as

$$\mathbf{S}_{\text{afe}} = \frac{2}{3}(S_1 + \mathbf{a}S_2 + \mathbf{a}^2 S_3) \tag{6.10}$$

where S_1, S_2, and S_3 are the switching states of each rectifier leg, as shown in Figure 6.3, and take the value of 0 if S_x is off, or 1 if S_x is on ($x = 1, 2, 3$).

The input current dynamics equation (6.5) can be rewritten in the stationary $\alpha\beta$ frame as the following vector equation:

$$L_s \frac{d\mathbf{i}_s}{dt} = \mathbf{v}_s - \mathbf{v}_{\text{afe}} - R_s \mathbf{i}_s \tag{6.11}$$

where $\mathbf{i}_s$ is the input current vector, $\mathbf{v}_s$ is the supply line voltage, and $\mathbf{v}_{\text{afe}}$ is the voltage generated by the converter.

6.2.2 Discrete-Time Model

The predicted current is calculated using the discrete-time equation

$$\mathbf{i}_s(k+1) = \left(1 - \frac{R_s T_s}{L_s}\right)\mathbf{i}_s(k) + \frac{T_s}{L_s}[\mathbf{v}_s(k) - \mathbf{v}_{\text{afe}}(k)] \tag{6.12}$$

obtained from discretizing (6.11) for a sampling time T_s. The discretization is done by approximating the derivative as the difference over one sampling period as considered in the previous chapters and explained in Chapter 4.

Considering the input voltage and current vectors in orthogonal coordinates, the predicted instantaneous input active and reactive power can be expressed by the following equations:

$$P_{\text{in}}(k+1) = Re\{\mathbf{v}_s(k+1)\bar{\mathbf{i}}_s(k+1)\} = v_{s\alpha}i_{s\alpha} + v_{s\beta}i_{s\beta} \tag{6.13}$$

$$Q_{\text{in}}(k+1) = Im\{\mathbf{v}_s(k+1)\bar{\mathbf{i}}_s(k+1)\} = v_{s\beta}i_{s\alpha} - v_{s\alpha}i_{s\beta} \tag{6.14}$$

where $\bar{\mathbf{i}}_s(k+1)$ is the complex conjugate of the predicted input current vector $\mathbf{i}_s(k+1)$, for a given voltage vector generated by the rectifier $\mathbf{v}_{\text{afe}}$.

For a small sampling time, with respect to the grid fundamental frequency, it can be assumed that $\mathbf{v}_s(k+1) \approx \mathbf{v}_s(k)$. However, if the sampling time is not small enough to consider the grid voltage as constant between two sampling intervals, the future grid voltage $\mathbf{v}_s(k+1)$ can be calculated by compensating the angle of the voltage vector for one sampling time:

$$\mathbf{v}_s(k+1) = \mathbf{v}_s(k)e^{j\Delta\theta} \tag{6.15}$$

where $\Delta\theta = \omega T_s$ is the angle advance of the grid voltage vector in one sampling interval and ω is the angular frequency of the grid voltage.

6.3 Predictive Current Control in an Active Front-End

The control scheme for the VOC of an AFE using predictive control for the current control loop is shown in Figure 6.4. A PI controller is used for DC link voltage regulation and generates the amplitude of the input current reference. The reference current is calculated by multiplying the output of the PI controller by the grid voltage waveform.

6.3.1 Cost Function

The predictive current controller must achieve the smallest current error with fast dynamics. An appropriate cost function is a measure of the predicted input current error. The following cost function considers the absolute error between the reference current and the predicted current, expressed in orthogonal coordinates:

$$g = |i^*_{s\alpha} - i^p_{s\alpha}| + |i^*_{s\beta} - i^p_{s\beta}| \tag{6.16}$$

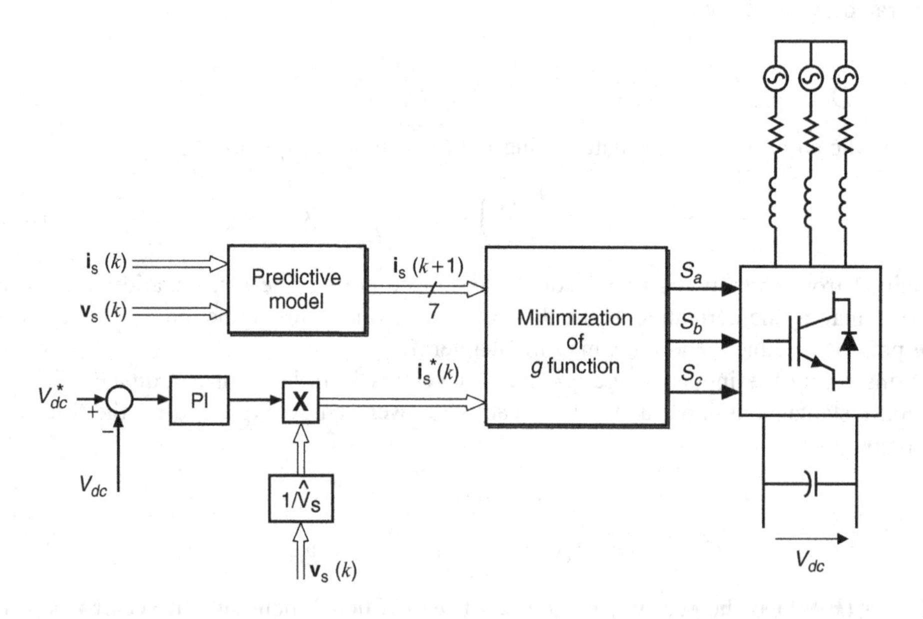

Figure 6.4 VOC of an AFE using predictive current control

Table 6.1 System parameters

Parameter	Value
V_{dc}^*	500 V
L_s	10 mH
R_s	0.1 Ω
v_s	150 V_{peak}
T_s	50 μs

where $\mathbf{i}_s(k+1) = i_{s\alpha}^p + ji_{s\beta}^p$ is the input current prediction at time $k+1$ for a given converter voltage $\mathbf{v}_{\text{afe}}(k)$, and the current $\mathbf{i}_s^*(k) = i_{s\alpha}^* + ji_{s\beta}^*$ corresponds to the reference current at time k.

Results for the system shown in Figure 6.4 were obtained using the parameters shown in Table 6.1. Results at steady state and the behavior of the system during transients are presented here.

The input current and voltage, and the converter voltage at steady state operation, are shown in Figure 6.5. Here, the input current is sinusoidal and in phase with the supply voltage. It can be observed that the predictive controller generates a modulated voltage without using a modulator.

The behavior of the predictive current control of an AFE for a load step, from half to full load, is presented in Figure 6.6. Here, the fast dynamic response of the input current control allows for fast compensation of the load step. The input current is in phase with the supply voltage even during the transient, as shown in Figure 6.7.

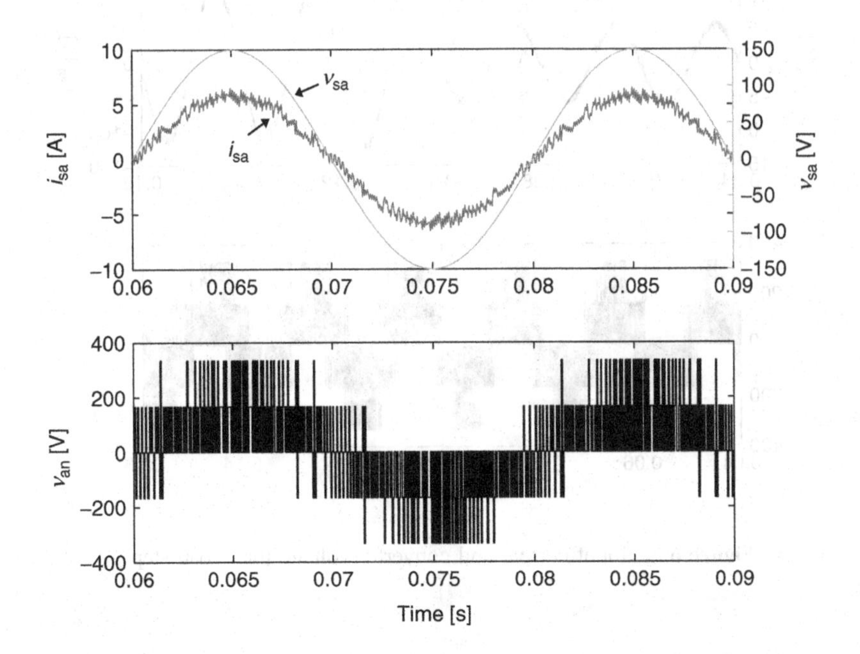

Figure 6.5 Input current and converter voltage in steady state operation

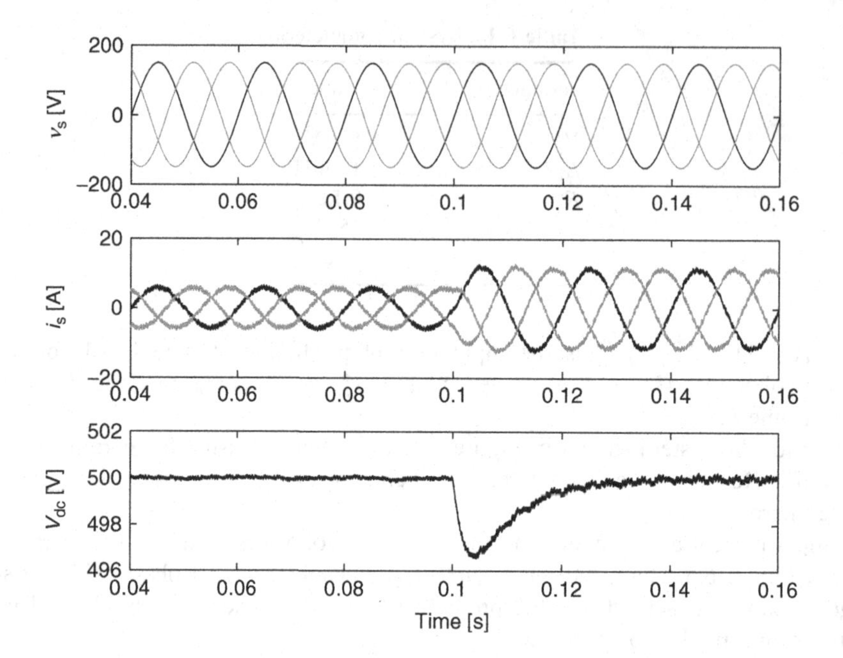

Figure 6.6 Supply voltage, input current, and DC link voltage for a load step

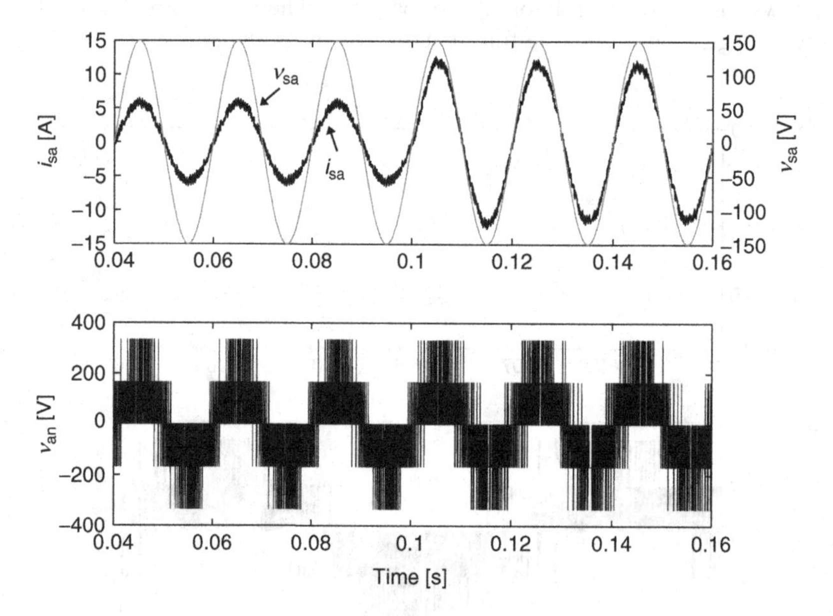

Figure 6.7 Input current and converter voltage for a load step

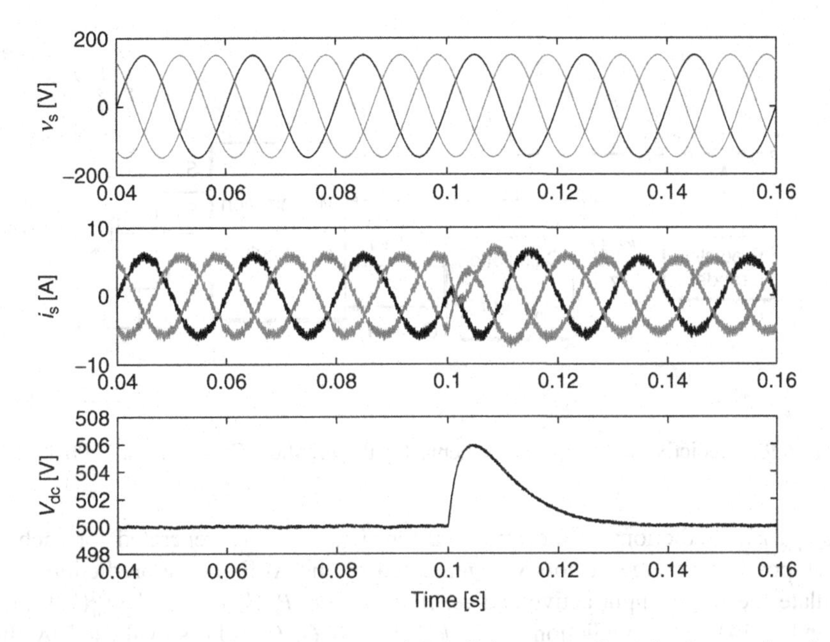

Figure 6.8 Supply voltage, input current, and DC link voltage for a load step changing from rectifying to regenerative operation

A change from rectifying to regenerative operation was performed. The supply voltage, input current, and DC link voltage during this test are shown in Figure 6.8. A step change in the direction of the DC load current is performed at time 0.1 s, and produces a momentary rise in the DC link voltage. This change in the DC link voltage is compensated by the voltage controller resulting in a change in the sign of the reference currents. It can be observed that, at time 0.1 s, the input current is shifted by 180° with respect to the supply voltage and the unity power factor is maintained.

6.4 Predictive Power Control

By using the rectifier model and considering the instantaneous power theory [7], it is possible to predict the behavior of the input active power and reactive power at the input of the converter. Then, by defining an appropriate cost function it is possible to directly control the power flow between the converter and the grid [8].

The switching state of the converter is changed at equidistant time instants and is constant during a whole sampling interval. On each sampling interval, the control strategy selects the switching state that will be applied by minimization of a cost function.

Predictive power control has no internal control loops and does not need external modulators. The currents are forced by directly controlling the active and reactive power.

6.4.1 Cost Function and Control Scheme

The block diagram of the control strategy is shown in Figure 6.9. The input currents $\mathbf{i}_s(k)$ are measured and the future current $\mathbf{i}_s(k+1)$ is calculated using the applied converter

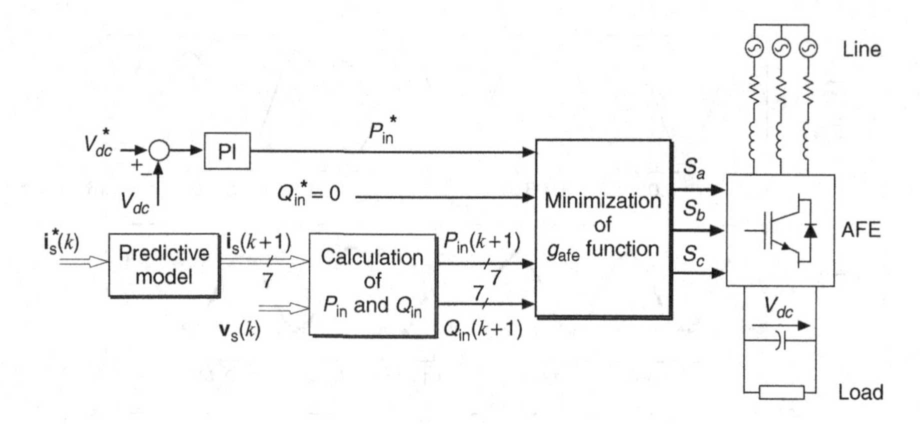

Figure 6.9 Predictive power control scheme for the rectifier (Cortes *et al.*, 2008 © IEEE)

voltage $\mathbf{v}_{\text{afe}}(k)$. Predictions of the future current $\mathbf{i}_s(k+1)$ are generated for each one of the seven possible voltage vectors $\mathbf{v}_{\text{afe}}$ generated by the AFE. These predictions are used to calculate the future input active and reactive power, $P_{\text{in}}(k+1)$ and $Q_{\text{in}}(k+1)$, using (6.13) and (6.14). Each prediction of $P_{\text{in}}(k+1)$ and $Q_{\text{in}}(k+1)$ is evaluated by the cost function g_{afe}.

The cost function g_{afe} summarizes the desired behavior of the rectifier: minimize the reactive power error, and control the active power P_{in} to be equal to a reference value P_{in}^*:

$$g_{\text{afe}} = |Q_{\text{in}}^* - Q_{\text{in}}(k+1)| + |P_{\text{in}}^* - P_{\text{in}}(k+1)| \tag{6.17}$$

A PI controller is used for DC link voltage regulation. The output of the PI corresponds to the power needed to compensate the error in the DC link voltage. This variable has been designated as the active power reference P_{in}^*.

The reference value for the reactive power Q_{in}^* is usually set to zero, as shown in Figure 6.9. However, in some applications it can have other values different than zero.

Each possible voltage vector $\mathbf{v}_{\text{afe}}(k)$ will generate a different value of the cost function g_{afe}. The voltage vector that minimizes the g_{afe} function, that is, the error in the input power, will be selected and applied during the next sampling interval.

The behavior of the predictive power control scheme is shown next, simulated for a 5 kW converter and a sampling time $T_s = 50\,\mu\text{s}$ for the control. The parameters of the system used for the simulation are: $v_s = 150\,\text{V}_{peak}$, $L_s = 10\,\text{mH}$, $R_s = 100\,\text{m}\Omega$, $C = 470\,\mu\text{F}$.

The performance of the power control for the rectifier was tested by applying a step in the active power reference P_{in}^*. The external PI controller used for voltage control is disconnected during this test. As shown in Figure 6.10a, the power tracking is very fast and there is no coupling between active and reactive power. The input currents, shown in Figure 6.10b, are sinusoidal and in phase with the line voltages.

Results for the predictive direct power control inside a voltage control loop are shown in Figure 6.11 for a step in the DC load. The active power reference, generated by the voltage controller, is followed with fast dynamics and without affecting the reactive power. The input currents for this test are also shown.

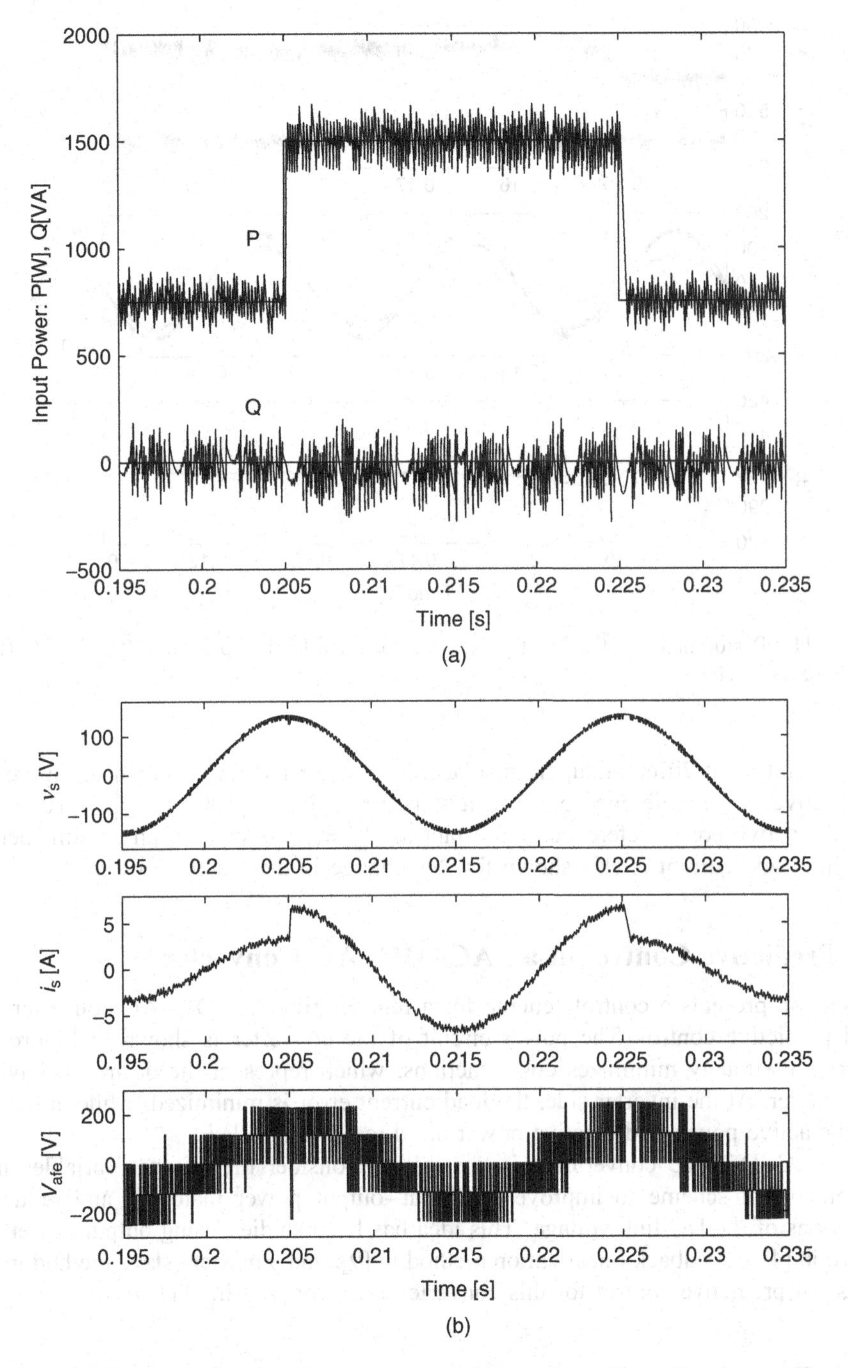

Figure 6.10 Behavior of the predictive power control for a step in the active power reference P_{in}^* from 750 to 1500 W. (a) Active and reactive power. (b) Input voltage, input current, and converter voltage (Cortes *et al.*, 2008 © IEEE)

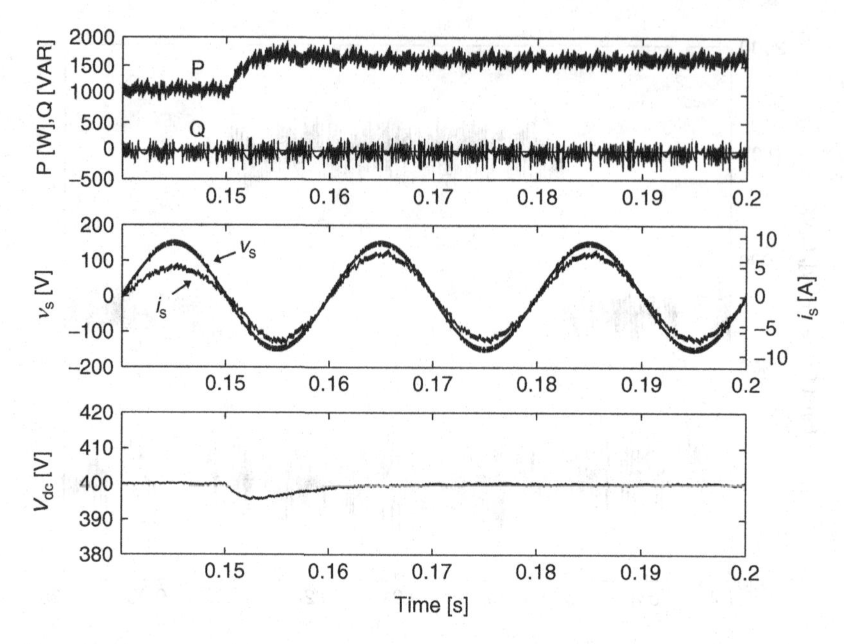

Figure 6.11 Performance of the input power control for a load step from 150 to 100 Ω (Cortes *et al.*, 2008 © IEEE)

The operation at different supply displacement power factors is shown in Figure 6.12. The reactive power reference presents a step from $Q_{in}^* = -1000$ to $Q_{in}^* = 1000$ VAR, while the active power reference is constant at $P_{in}^* = 1500$ W. The phase shift between the voltage and current is also shown for this change in the reactive power.

6.5 Predictive Control of an AC–DC–AC Converter

This section presents a control scheme for a regenerative AC–DC–AC converter using model predictive control. The power circuit of the converter is shown in Figure 6.13. The control strategy minimizes cost functions, which represent the desired behavior of the converter. At the inverter side, the load current error is minimized, while at the input side, the active power and reactive power are directly controlled.

In an AC–DC–AC converter it is possible to consider the inverter variables in the rectifier control scheme to improve the input–output power matching and reduce the fluctuations of the DC link voltage. This idea has been studied using output power feed-forward in [4], a feedback linearization method in [9], and a master–slave method in [10]. The use of predictive control for this converter was proposed in [11].

6.5.1 Control of the Inverter Side

The control of the inverter side is similar to the scheme presented in Chapter 4. The effect of each possible voltage vector generated by the inverter on the behavior of the load current is predicted using the load model. Then, each prediction is evaluated using a

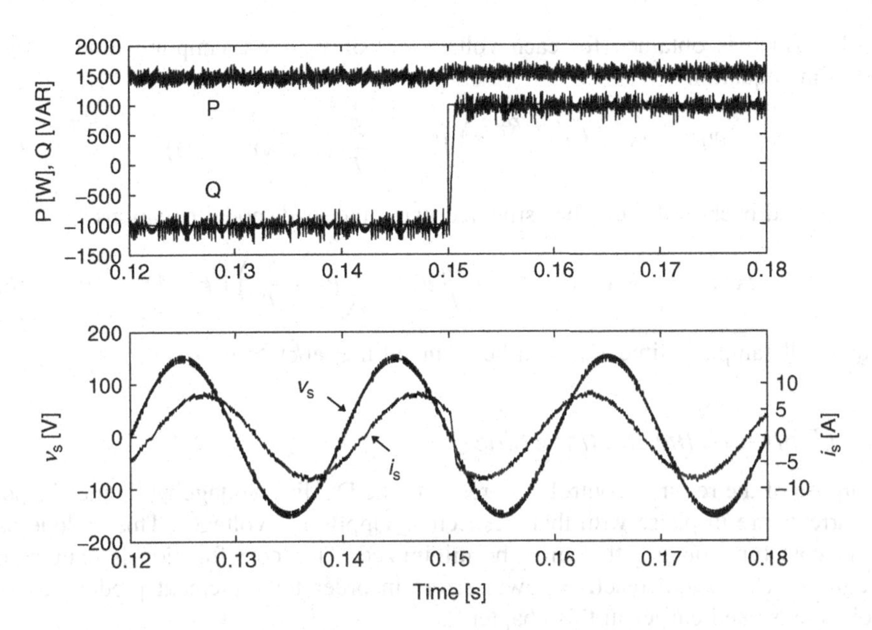

Figure 6.12 Operation at different supply power displacement factors. Step in the reactive power reference from $Q_{in}^* = -1000$ to $Q_{in}^* = 1000$ VAR (Cortes *et al.*, 2008 © IEEE)

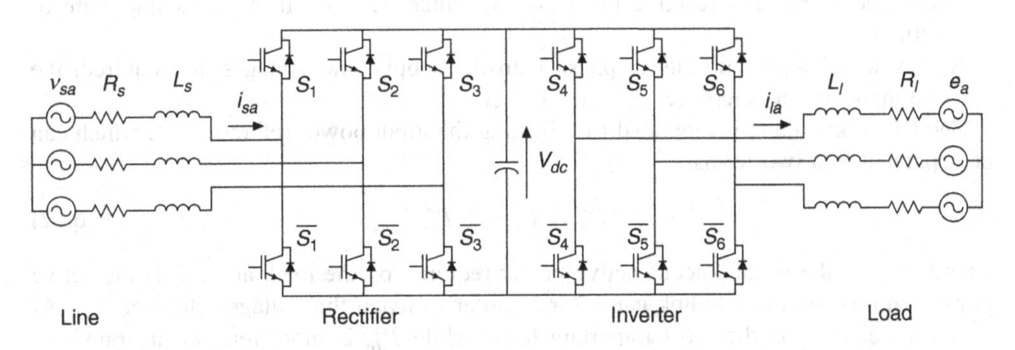

Figure 6.13 AC–DC–AC converter model (Rodriguez *et al.*, 2005 © IEEE)

cost function g_{inv}. The voltage vector that minimizes this function is selected and applied during the next sampling period. The cost function to be minimized for the inverter is the error between the reference and predicted current, which is expressed in orthogonal coordinates as

$$g_{inv} = |i_{l\alpha}^* - i_{l\alpha}^p| + |i_{l\beta}^* - i_{l\beta}^p| \tag{6.18}$$

where $i_{l\alpha}^*$ and $i_{l\beta}^*$ are the real and imaginary parts of the reference load current vector $\mathbf{i}_l^*$. Variables $i_{l\alpha}^p$ and $i_{l\beta}^p$ are the real and imaginary parts of the predicted load current vector

$\mathbf{i}_l^{\mathrm{p}}(k+1)$, which is obtained for each voltage vector $\mathbf{v}_{\mathrm{inv}}$ and sampling time T_s by the discrete-time equation

$$\mathbf{i}_l^{\mathrm{p}}(k+1) = \left(1 - \frac{R_l T_s}{L_l}\right)\mathbf{i}_l(k) + \frac{T_s}{L}\left(\mathbf{v}_{\mathrm{inv}}(k) - \mathbf{e}(k)\right) \tag{6.19}$$

where the load back-emf $\mathbf{e}$ can be estimated using the load model equations

$$\hat{\mathbf{e}}(k-1) = \mathbf{v}_{\mathrm{inv}}(k-1) - \frac{L_l}{T_s}\mathbf{i}(k) - \left(R_l - \frac{L_l}{T_s}\right)\mathbf{i}(k-1) \tag{6.20}$$

and for small sampling time T_s it can be assumed that $\mathbf{e}(k) \approx \hat{\mathbf{e}}(k-1)$.

6.5.2 Control of the Rectifier Side

The purpose of the rectifier control is to regulate the DC link voltage while the sinusoidal input currents are in phase with their respective supply line voltages. This is done using a proper cost function g_{afe} that must be minimized. This cost function is expressed as a function of active and reactive power errors in order to implement predictive power control as presented earlier in this chapter:

$$g_{\mathrm{afe}} = |Q_{\mathrm{in}}^* - Q_{\mathrm{in}}| + |P_{\mathrm{in}}^* - P_{\mathrm{in}}| \tag{6.21}$$

where Q_{in}^* and P_{in}^* are the required active and reactive input power, and Q_{in} and P_{in} are the predicted active and reactive input power, which depend on the switching state of the rectifier.

As sinusoidal input currents in phase with the supply line voltages are required, the reactive input power reference Q_{in}^* must be zero.

The DC link voltage is regulated by adjusting the input power reference P_{in}^*, which can be separated into two terms:

$$P_{\mathrm{in}}^* = P_{load}^* + P_{dc}^* \tag{6.22}$$

where P_{load}^* is the instantaneous active power required by the load, and P_{dc}^* is the active power required by the DC link capacitor in order to reach the voltage reference V_{dc}^*. At steady state, P_{load}^* is the most important term, while P_{dc}^* is more relevant in transients and for compensation of unmodeled losses.

6.5.3 Control Scheme

The block diagram of the control strategy is shown in Figure 6.14. The input current $\mathbf{i}_s(k)$ is measured and predictions of the future current $\mathbf{i}_s(k+1)$ are generated for each one of the seven possible voltage vectors $\mathbf{v}_{\mathrm{afe}}$ generated by the AFE. These currents are calculated using the discrete-time equation (6.12).

These predictions are used to calculate the future input active and reactive power, $P_{\mathrm{in}}(k+1)$ and $Q_{\mathrm{in}}(k+1)$. These values are calculated using (6.13) and (6.14).

In order to calculate the active power reference P_{in}^*, an estimate of the load active power is needed. This value can be obtained using the load current reference and the

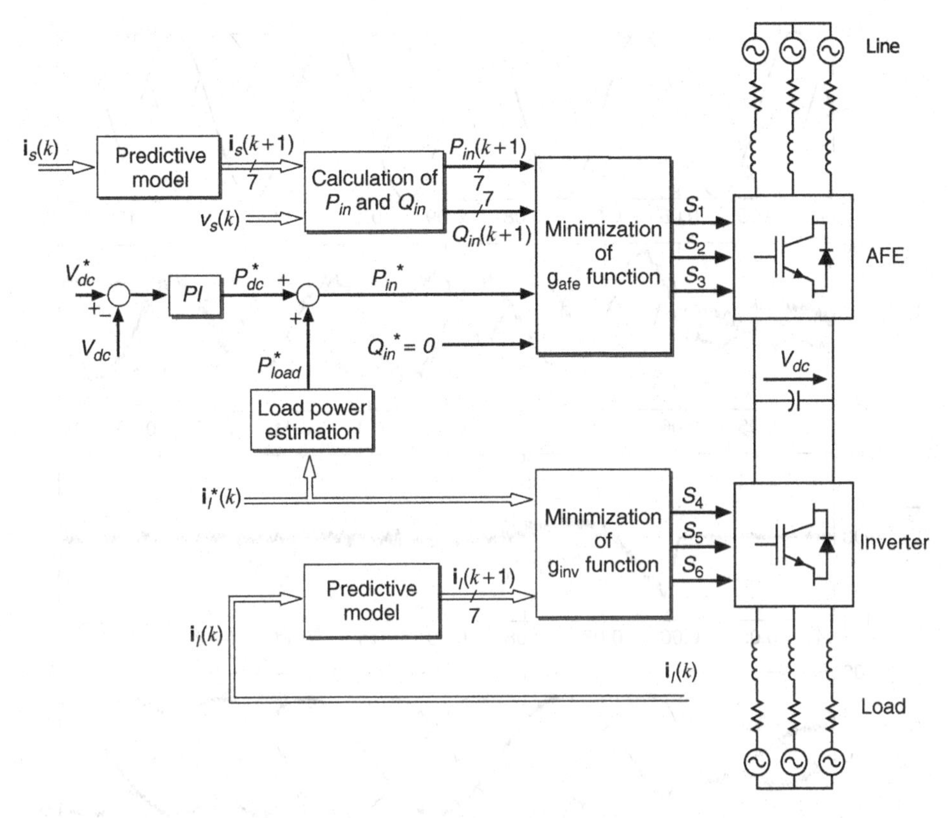

Figure 6.14 Predictive control scheme for the AC–DC–AC converter (Rodriguez *et al.*, 2005 © IEEE)

estimated value of the back-emf through the following expression:

$$P_{load}^* = R_l \mathbf{i}_l^* \bar{\mathbf{i}}_l^* + Re\{\mathbf{e}\bar{\mathbf{i}}_l^*\} \tag{6.23}$$

A PI controller is used for regulating the DC link voltage. The output of the PI corresponds to the power needed to compensate the error in the DC link voltage. This variable is designated as P_{dc}^*.

The behavior of this control scheme for a step in the amplitude of the load current reference is shown in Figure 6.15. The amplitude of the load current is changed from 35 to 70 A at time 0.06 s. The inverter currents respond with fast dynamics to this change causing a drop in the DC link voltage, which is compensated by the DC link controller by adjusting the active power reference. The input currents also present a fast dynamic behavior while the currents are kept sinusoidal and in phase with the supply voltages, even during the transient.

The behavior of the input active and reactive power for the same test is shown in Figure 6.16. The active power responds with fast dynamics to the reference change while the reactive power is not affected by the transient. The load power P_{load} and the output of the PI controller P_{dc} are also shown in this figure.

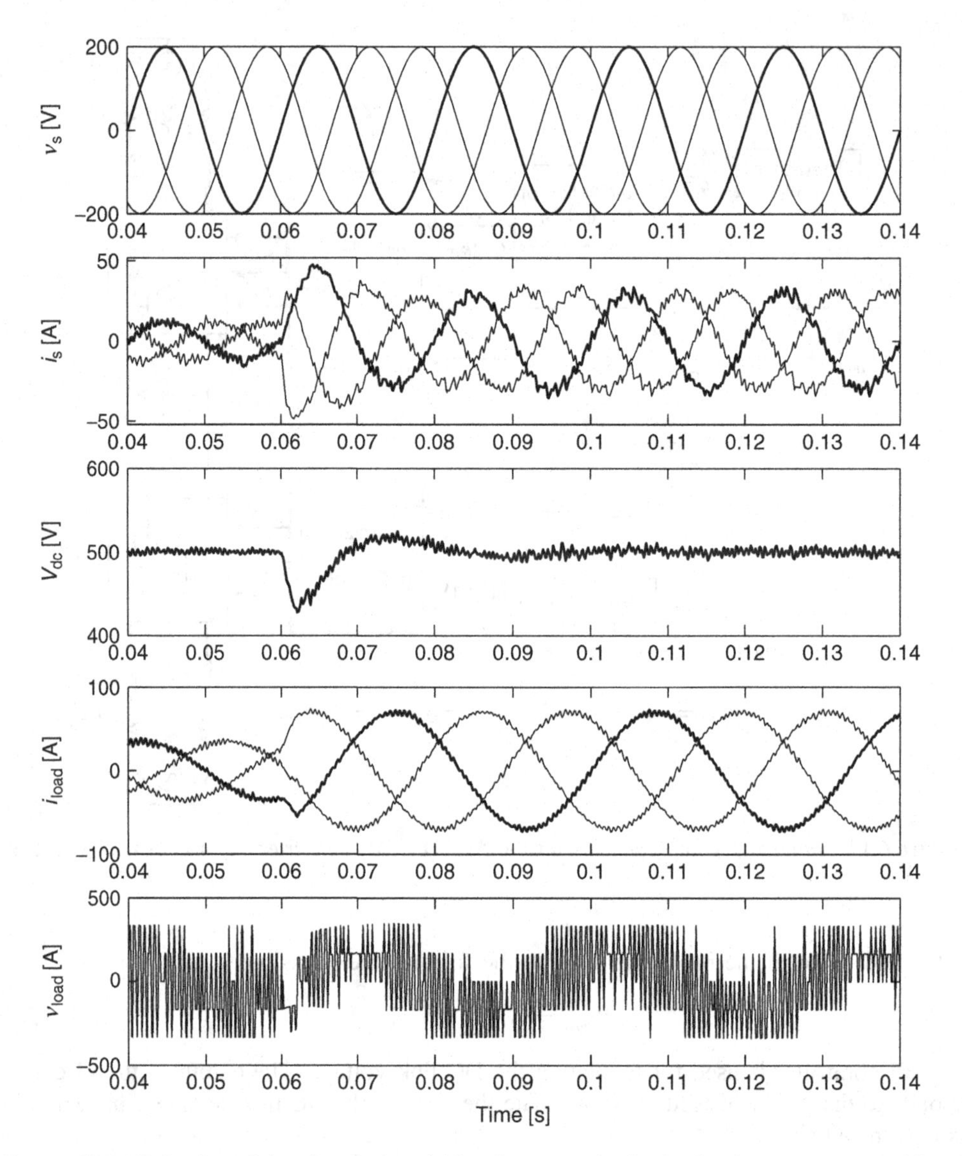

Figure 6.15 Behavior of the electrical variables for a step in the load reference current with an output frequency of 30 Hz (Rodriguez *et al.*, 2005 © IEEE)

6.6 Summary

Two predictive control schemes are presented in this chapter. First a voltage-oriented control scheme is considered, using a predictive control strategy for the grid currents. The second control scheme is based on the idea of direct power control, using the active and reactive power error in the cost function.

Both control schemes consist of an inner control loop, for currents or power, and an outer voltage control loop for the DC link. A future challenge is to include the control of

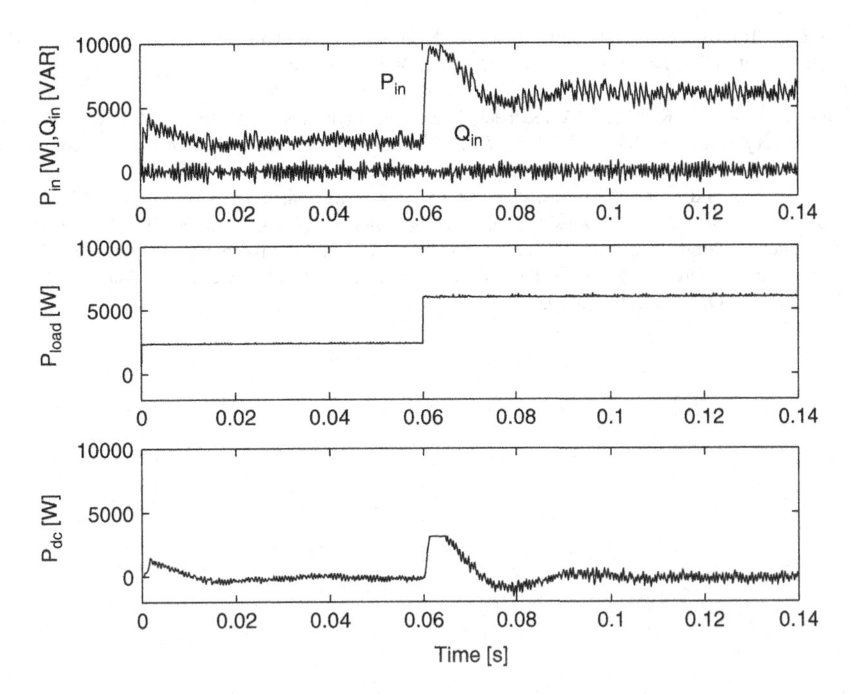

Figure 6.16 Input, output, and DC link power for a step in the load reference current (Rodriguez *et al.*, 2005 © IEEE)

the DC link voltage inside the predictive control scheme, avoiding the use of a cascaded structure.

The use of predictive power control in a regenerative drive application is also described in this chapter. This application considers a predictive current controller for the inverter side and a feedforward loop for rectifier control.

References

[1] N. Mohan, T. Underland, and W. Robbins, *Power Electronics*, 2nd ed. John Wiley & Sons, Inc. 1995.

[2] J. Rodríguez, J. Dixon, J. Espinoza, J. Pont, and P. Lezana, "PWM regenerative rectifiers: state of the art," *IEEE Transactions on Industrial Electronics*, vol. 52, no. 1, pp. 5–22, February 2005.

[3] M. Malinowski, M. P. Kazmierkowski, and A. M. Trzynadlowski, "A comparative study of control techniques for PWM rectifiers in AC adjustable speed drives," *IEEE Transactions on Power Electronics*, vol. 18, no. 6, pp. 1390–1396, November 2003.

[4] T. G. Habetler, "A space vector-based rectifier regulator for AC/DC/AC converters," *IEEE Transactions on Power Electronics*, vol. 8, no. 1, pp. 30–36, January 1993.

[5] T. Noguchi, H. Tomiki, S. Kondo, and I. Takahashi, "Direct power control of PWM converter without power-source voltage sensors," *IEEE Transactions on Industry Applications*, vol. 34, no. 3, pp. 473–479, May/June 1998.

[6] M. Malinowski, M. Jasinski, and M. P. Kazmierkowski, "Simple direct power control of three-phase PWM rectifier using space-vector modulation (DPC-SVM)," *IEEE Transactions on Industrial Electronics*, vol. 51, no. 2, pp. 447–454, April 2004.

[7] H. Akagi, E. Watanabe, and M. Aredes, *Instantaneous Power Theory and Applications to Power Conditioning*, IEEE Press Series on Power Engineering. John Wiley & Sons, Inc. 2007.

[8] P. Cortés, J. Rodríguez, P. Antoniewicz, and M. Kazmierkowski, "Direct power control of an AFE using predictive control," *IEEE Transactions on Power Electronics*, vol. 23, no. 5, pp. 2516–2523, September 2008.
[9] J. Jung, S. Lim, and K. Nam, "A feedback linearizing control scheme for a PWM converter-inverter having a very small DC-link capacitor," *IEEE Transactions on Industry Applications*, vol. 35, no. 5, pp. 1124–1131, September/October 1999.
[10] N. Hur, J. Jung, and K. Nam, "A fast dynamic DC-link power-balancing scheme for a PWM converter-inverter system," *IEEE Transactions on Industrial Electronics*, vol. 48, no. 4, pp. 794–803, August 2001.
[11] J. Rodríguez, J. Pontt, P. Correa, P. Lezana, and P. Cortés, "Predictive power control of an AC/DC/AC converter," in Conference Record of the Industry Applications Conference, 2005. Fourtieth IAS Annual Meeting, vol. 2, October 2005, pp. 934–939.

7

Control of a Matrix Converter

7.1 Introduction

The matrix converter (MC) is a single-stage power converter, capable of feeding directly a m-phase load from a n-phase source ($n \times m$ MC) without energy storage devices [1].

The most relevant features of a MC are:

1. The power circuit is compact.
2. It delivers voltages and currents to the load with high quality and without restriction on the frequency.
3. It can generate sinusoidal input current and operate with unity power factor.
4. It allows power to flow from the source to the load and in the opposite direction. This means it is very suitable for regenerative loads.

These are the characteristics of an ideal converter and this is the reason for the great interest in this topology, which has been intensively studied for approximately three decades, starting with the pioneering work of Venturini and Alesina [2, 3]. This chapter will explain the working principle of the MC and will introduce the application of predictive control to control the waveform of the load and input currents.

7.2 System Model

7.2.1 *Matrix Converter Model*

The power circuit of the MC is presented in Figure 7.1. It uses a set of bidirectional switches to directly connect the three-phase power supply to a three-phase load. This is a 3×3 MC. As shown in Figure 7.1, each bidirectional switch is composed of two power transistors with their parallel diodes in anti-series connection.

The MC is connected to the three-phase source through the input filter L_f, R_f, C_f. This filter has two main purposes:

1. To avoid the generation of overvoltages, produced by the short-circuit impedance of the power supply (not shown in the figure), due to the fast commutation of currents i_{eu}, i_{ev}, and i_{ew}.
2. To eliminate high-frequency harmonics in the input currents i_u, i_v, i_w.

Predictive Control of Power Converters and Electrical Drives, First Edition. Jose Rodriguez and Patricio Cortes.

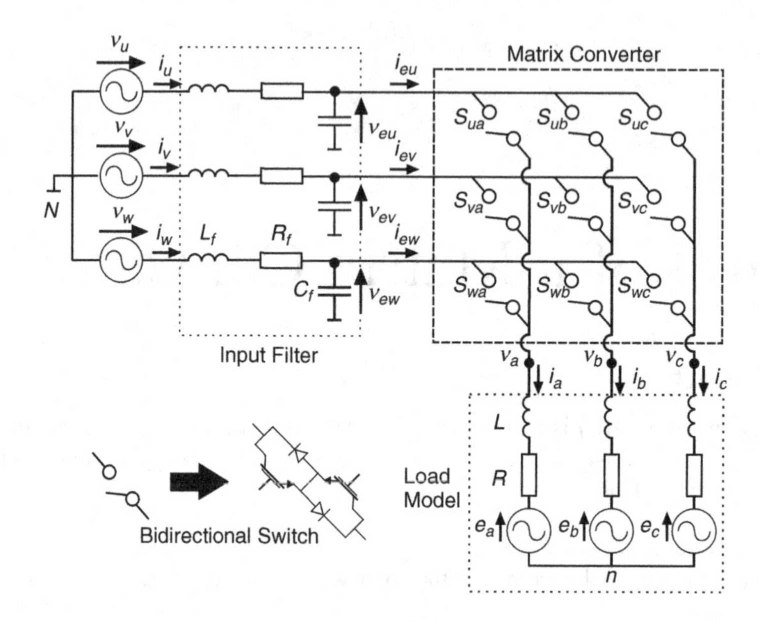

Figure 7.1 Power circuit of the MC

In Figure 7.1 each bidirectional switch is associated with a variable defined as S_{xy} with $x \in \{u, v, w\}$ and $y \in \{a, b, c\}$. The conduction state of each bidirectional switch is determined exclusively by the value of its control signal. S_{xy} is also known as the switching function for switch xy. $S_{xy} = 1$ implies that switch xy is on, closed, or conducting, while $S_{xy} = 0$ means that the switch is off, open, or in blocking state.

It must be mentioned that the load current must not be interrupted abruptly, because the inductive nature of the load will generate an important overvoltage that can destroy the components. In addition, operation of the switches cannot short-circuit two input lines, because this switching state will originate short-circuit currents. These restrictions can be expressed in mathematical form by the following equation:

$$S_{uy} + S_{vy} + S_{wy} = 1 \quad \forall \quad y \in \{a, b, c\} \tag{7.1}$$

Referenced to the neutral point N, the relation between the load and input voltages of the MC is expressed as

$$\begin{bmatrix} v_a(t) \\ v_b(t) \\ v_c(t) \end{bmatrix} = \underbrace{\begin{bmatrix} S_{ua} & S_{va} & S_{wa} \\ S_{ub} & S_{vb} & S_{wb} \\ S_{uc} & S_{vc} & S_{wc} \end{bmatrix}}_{T} \cdot \begin{bmatrix} v_{eu}(t) \\ v_{ev}(t) \\ v_{ew}(t) \end{bmatrix} \tag{7.2}$$

where T is the instantaneous transfer matrix.

The input and load voltages can be expressed as vectors as follows:

$$\mathbf{v}_o = \begin{bmatrix} v_a(t) \\ v_b(t) \\ v_c(t) \end{bmatrix} \qquad \mathbf{v}_i = \begin{bmatrix} v_{eu}(t) \\ v_{ev}(t) \\ v_{ew}(t) \end{bmatrix} \tag{7.3}$$

Using the definitions of (7.3), the relation of the voltages is given by

$$\mathbf{v}_o = T \cdot \mathbf{v}_i \tag{7.4}$$

Applying Kirchhoff's current law to the switches, the following equation can be obtained:

$$\begin{bmatrix} i_{eu}(t) \\ i_{ev}(t) \\ i_{ew}(t) \end{bmatrix} = \underbrace{\begin{bmatrix} S_{ua} & S_{ub} & S_{uc} \\ S_{va} & S_{vb} & S_{vc} \\ S_{wa} & S_{wb} & S_{wc} \end{bmatrix}}_{T^T} \cdot \begin{bmatrix} i_a(t) \\ i_b(t) \\ i_c(t) \end{bmatrix} \tag{7.5}$$

Considering the current vectors

$$\mathbf{i}_i = \begin{bmatrix} i_{eu}(t) \\ i_{ev}(t) \\ i_{ew}(t) \end{bmatrix} \qquad \mathbf{i}_o = \begin{bmatrix} i_a(t) \\ i_b(t) \\ i_c(t) \end{bmatrix} \tag{7.6}$$

the equation for the current is

$$\mathbf{i}_i = T^T \cdot \mathbf{i}_o \tag{7.7}$$

where T^T is the transpose of matrix T.

7.2.2 Working Principle of the Matrix Converter

The bidirectional switches open and close, operating with a high switching frequency, to generate a low-frequency voltage with variable amplitude and frequency. This goal is achieved by generating switching patterns as shown in Figure 7.2. The low-frequency component of the load voltage is synthesized by sampling the input voltages closing and opening the bidirectional switches.

If we define t_{ij} as the time during which switch S_{ij} is closed (on) and T is the sampling interval, we can express the low-frequency component of the load voltage as

$$\bar{v}_{jN}(t) = \frac{t_{uj} \cdot v_{eu}(t) + t_{vj} \cdot v_{ev}(t) + t_{wj} \cdot v_{ew}(t)}{T} \qquad j \in \{a, b, c\} \tag{7.8}$$

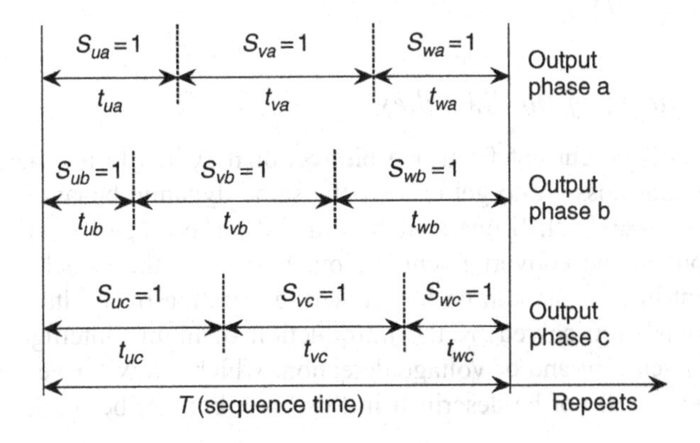

Figure 7.2 Switching patterns for the MC

where $\bar{v}_{jN}(t)$ is the low-frequency component (mean value calculated over one sampling interval T) of output phase j.

From Figure 7.2 it can observed, for example, that the voltage of phase a is generated by delivering load voltage v_{eu} during time t_{ua}, voltage v_{ev} during time t_{va}, and voltage v_{we} during time t_{wa}.

The conduction times must comply with the restriction

$$T = t_{uj} + t_{vj} + t_{wj} \qquad \forall j \in \{\mathrm{a, b, c}\} \tag{7.9}$$

By defining the duty cycles as

$$m_{uj}(t) = \frac{t_{uj}}{T}, m_{vj}(t) = \frac{t_{vj}}{T}, m_{wj}(t) = \frac{t_{wj}}{T} \tag{7.10}$$

and expanding (7.8) for each phase, the following equations can be obtained:

$$\begin{bmatrix} \bar{v}_{aN}(t) \\ \bar{v}_{bN}(t) \\ \bar{v}_{cN}(t) \end{bmatrix} = \underbrace{\begin{bmatrix} m_{ua}(t) & m_{va}(t) & m_{wa}(t) \\ m_{ub}(t) & m_{vb}(t) & m_{wb}(t) \\ m_{uc}(t) & m_{vc}(t) & m_{wc}(t) \end{bmatrix}}_{M(t)} \cdot \begin{bmatrix} v_{eu} \\ v_{ev} \\ v_{ew} \end{bmatrix} \tag{7.11}$$

$$\bar{\mathbf{v}}_o = \begin{bmatrix} \bar{v}_{aN} \\ \bar{v}_{bN} \\ \bar{v}_{cN} \end{bmatrix} \tag{7.12}$$

$$\bar{\mathbf{v}}_o(t) = M(t) \cdot \mathbf{v}_i(t) \tag{7.13}$$

where $\bar{\mathbf{v}}_o(t)$ is the low-frequency output voltage vector, $\bar{\mathbf{v}}_i(t)$ is the instantaneous input voltage vector, and $M(t)$ is the low-frequency transfer matrix of the MC.

By considering an analogous procedure for the input currents, it can be demonstrated that

$$\bar{\mathbf{i}}_i(t) = M^T(t) \cdot \mathbf{i}_o(t) \tag{7.14}$$

where $\bar{\mathbf{i}}_i(t)$ is the low-frequency component of the input current vector and $M^T(t)$ is the transpose of matrix $M(t)$.

7.2.3 Commutation of the Switches

The commutation of the current from one bidirectional switch to another is not an easy task, because it is not possible to get exactly the same dynamic behavior when semiconductors are turned on and off. If one switch is turned on too rapidly, it can cause a short circuit at the input of the converter. On the other hand, if the switch is turned on too slowly, the current in the load can be interrupted, generating overvoltages.

This problem has been solved by the introduction of highly intelligent commutation strategies based on current and/or voltage detection, which allow for very safe commutation. These methods will not be described in this book but can be found in [1].

7.3 Classical Control: The Venturini Method

The sinusoidal voltages of the three-phase power supply can be expressed as

$$\mathbf{v}_i(t) = \begin{bmatrix} v_u \\ v_v \\ v_w \end{bmatrix} = \begin{bmatrix} V_i \cdot \cos(w_i t) \\ V_i \cdot \cos(w_i t + 2\pi/3) \\ V_i \cdot \cos(w_i t + 4\pi/3) \end{bmatrix} \quad (7.15)$$

The desired voltages generated at the load, which are the low-frequency components, can be expressed as

$$\bar{\mathbf{v}}_o(t) = \begin{bmatrix} \bar{v}_{aN}(t) \\ \bar{v}_{bN}(t) \\ \bar{v}_{cN}(t) \end{bmatrix} = \begin{bmatrix} V_o \cdot \cos(w_o t) \\ V_o \cdot \cos(w_o t + 2\pi/3) \\ V_o \cdot \cos(w_o t + 4\pi/3) \end{bmatrix} \quad (7.16)$$

Neglecting the presence of the input filter, the relation between the amplitude of the input voltage and the amplitude of the output voltage is

$$V_o = q \cdot V_i \quad (7.17)$$

where q is the voltage gain.

Considering that the typical load of a MC will have low-pass characteristics, the output current is

$$\mathbf{i}_o(t) = \begin{bmatrix} i_a(t) \\ i_b(t) \\ i_c(t) \end{bmatrix} = \begin{bmatrix} I_o \cdot \cos(w_o t + \phi) \\ I_o \cdot \cos(w_o t + 2\pi/3 + \phi) \\ I_o \cdot \cos(w_o t + 4\pi/3 + \phi) \end{bmatrix} \quad (7.18)$$

To operate with unity power factor, the switches must be controlled to generate the following input currents (the fundamental components):

$$\bar{\mathbf{i}}_i(t) = \begin{bmatrix} \bar{i}_u \\ \bar{i}_v \\ \bar{i}_w \end{bmatrix} = \begin{bmatrix} I_i \cdot \cos(w_i t) \\ I_i \cdot \cos(w_i t + 2\pi/3) \\ I_i \cdot \cos(w_i t + 4\pi/3) \end{bmatrix} \quad (7.19)$$

The MC does not store energy and, for this reason, the active power at the input (P_i) and at the output (P_o) must be equal at all times, as expressed by the equation

$$P_i = \frac{3V_i I_i}{2} = \frac{3qV_i I_o \cdot \cos(\phi)}{2} = P_o \quad (7.20)$$

With all these fundamental definitions, the task of the modulator is to find a low-frequency transfer matrix $M(t)$ such that (7.13) and (7.7) are satisfied, considering the restrictions given by (7.15)–(7.20).

The solution for the matrix $M(t)$ can be obtained from [1] and [3] and can be reduced to a very compact expression given by

$$m_{ij} = \left[\, 1 + 2v_{iN}(t) \cdot \bar{v}_{jN}(t)/V_i^2 \,\right] \quad (7.21)$$

where $i \in \{u, v, w\}$ and $j \in \{a, b, c\}$.

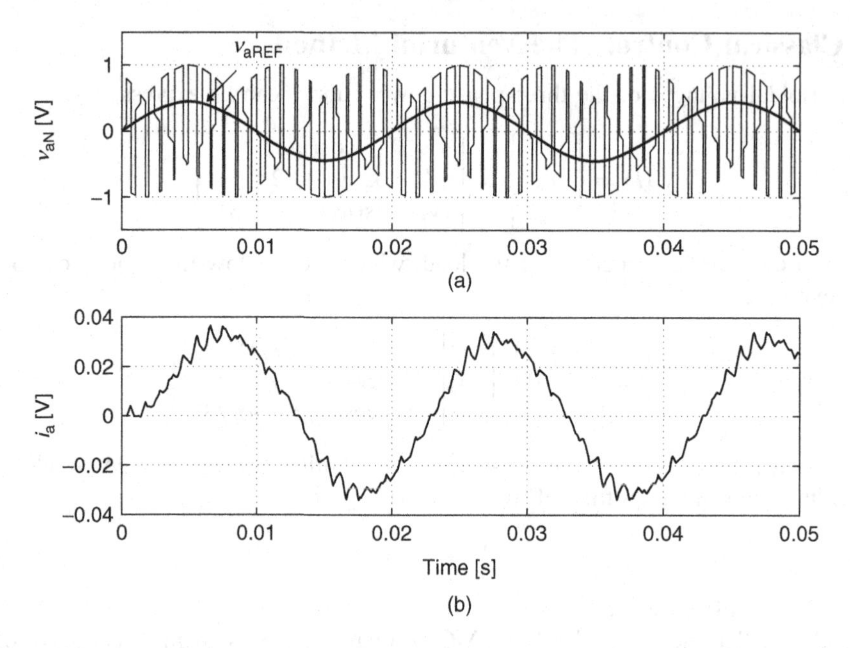

Figure 7.3 Output voltage v_{aN} and output current i_a generated by the Venturini method

The procedure for modulation of the MC can be resumed in the following steps:

1. Measure the grid voltages v_{iN} and the desired reference voltages $v_{jref} = \bar{v}_{jN}$.
2. Using (7.21), construct the low-frequency transfer matrix $M(t)$.
3. Calculate the conduction times for the nine switches t_{ij} using (7.10).
4. Generate the gate drive pulses for the bidirectional switches.

Figure 7.3 shows the operation of the MC controlled by the Venturini method. It can be observed that voltage v_{aN} is synthesized using all three phases of the grid. The load current i_a is very sinusoidal, with a small ripple that can be reduced even more by increasing the switching frequency.

Figure 7.4 presents the input voltage v_u and the input currents. It can be observed that the input current i_{eu} has strong commutation and that these abrupt changes are completely eliminated from i_u as a result of the filter action.

Space vector modulation can also be applied in MC. This technique will not be included in this book but more information about it can be found in [1] and [4].

7.4 Predictive Current Control of the Matrix Converter

7.4.1 Model of the Matrix Converter for Predictive Control

7.4.1.1 Matrix Converter Model

For predictive control the model of the MC is extremely simple: it considers only (7.2) and (7.5), which relate the instantaneous values of input and output currents and voltages.

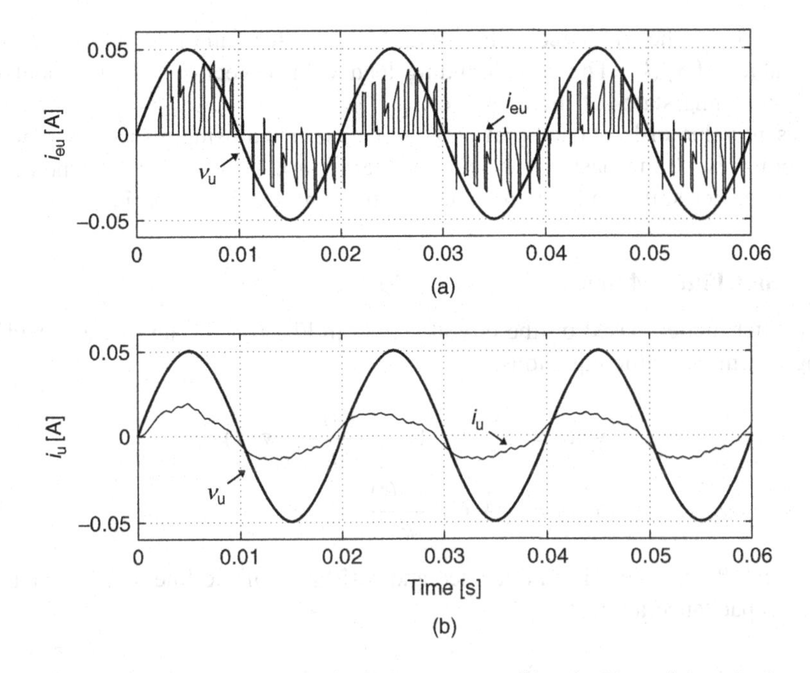

Figure 7.4 Input voltage and currents generated by the Venturini method

Based on the restriction presented by (7.1), the MC has 27 different switching states to be considered for prediction of the variables.

7.4.1.2 Load Model

In this case the objective is to obtain an equation to predict the value of the load current in the next sampling interval, for each of the 27 different switching states of the MC. The equation for the resistive–inductive–active load presented in Figure 7.1 is

$$L\frac{d\mathbf{i}_o(t)}{dt} = \mathbf{v}_o(t) - R\mathbf{i}_o(t) - \mathbf{e}(t) \tag{7.22}$$

where L and R are the inductance and resistance of the load and $\mathbf{e}$ is the electromotive force (emf). This load model is quite general, because it covers a wide variety of applications such as passive inductive load, motors, and grid-connected converters.

Considering the approximation for the derivative of the output current

$$\frac{d\mathbf{i}_o}{dt} \approx \frac{\mathbf{i}_o(k+1) - \mathbf{i}_o(k)}{T_s} \tag{7.23}$$

where T_s is the sampling period, the equation for predicting the load current is obtained from substituting (7.22) into (7.23), which gives

$$\mathbf{i}_o(k+1) = \left(1 - \frac{RT_s}{L}\right)\mathbf{i}_o(k) + \frac{T_s}{L}\left(\mathbf{v}_o(k) - \hat{\mathbf{e}}(k)\right) \tag{7.24}$$

where $\mathbf{i}_o(k+1)$ is the predicted value of the current for sampling interval $k+1$, for different values of $\mathbf{v}_o(k)$. The corresponding load voltage vector $\mathbf{v}_o(k)$ is calculated for all the 27 switching states of the converter.

The present value of load back-emf $\mathbf{e}(t)$ can be estimated using a second-order extrapolation from present and past values, or considering $e(k-1) \approx e(k)$, depending on the sampling time. For sufficiently small sampling time, no extrapolation is needed.

7.4.1.3 Input Filter Model

The input filter model, based on the circuit shown in Figure 7.1, can be described by the following continuous-time equations:

$$\mathbf{v}_s(t) = R_f \mathbf{i}_s(t) + L_f \frac{d\mathbf{i}_s(t)}{dt} + \mathbf{v}_i(t) \tag{7.25}$$

$$\mathbf{i}_s(t) = \mathbf{i}_i + C_f \frac{d\mathbf{v}_i(t)}{dt} \tag{7.26}$$

where L_f and R_f are the joint inductance and resistance of the line and filter and C_f is the filter's capacitance and:

$$\mathbf{v}_s(t) = 2/3(v_u + \mathbf{a}v_v + \mathbf{a}^2 v_w)$$

$$\mathbf{i}_s(t) = 2/3(i_u + \mathbf{a}i_v + \mathbf{a}^2 i_w)$$

This continuous-time system can be rewritten as

$$\dot{\mathbf{x}}(t) = \underbrace{\begin{bmatrix} 0 & 1/C_f \\ -1/Lf & -R_f/L_f \end{bmatrix}}_{\mathbf{A}_c} \mathbf{x}(t) + \underbrace{\begin{bmatrix} 0 & -1/C_f \\ 1/Lf & 0 \end{bmatrix}}_{\mathbf{B}_c} \mathbf{u}(t) \tag{7.27}$$

with

$$\mathbf{x}(t) = \begin{bmatrix} \mathbf{v}_i(t) \\ \mathbf{i}_s(t) \end{bmatrix} \quad \text{and} \quad \mathbf{u}(t) = \begin{bmatrix} \mathbf{v}_s(t) \\ \mathbf{i}_i(t) \end{bmatrix} \tag{7.28}$$

A discrete state space model can be derived when a zero-order hold input is applied to a continuous-time system described in state space form as in (7.27). Considering a sampling period T_s, the discrete-time system derived from (7.27) is

$$\mathbf{x}(k+1) = \mathbf{A}_q \mathbf{x}(k) + \mathbf{B}_q \mathbf{u}(k) \tag{7.29}$$

with

$$\mathbf{A}_q = e^{\mathbf{A}_c T_s} \quad \text{and} \quad \mathbf{B}_q = \int_0^{T_s} e^{\mathbf{A}_c (T_s - \tau)} \mathbf{B}_c \, d\tau \tag{7.30}$$

For more details on sampled-data systems and the theory employed in this analysis, see [5]. The discrete-time variables will match the continuous-time variables at sampling intervals. A convenient way of obtaining the discrete model is through MATLAB® function *c2d():conversion of continuous-time models to discrete time.* To predict the mains current, it is necessary simply to solve $\mathbf{i}_s(k+1)$ from (7.29):

$$\begin{aligned}\mathbf{i}_s(k+1) &= \mathbf{A}_q(2,1)\mathbf{v}_e(k) + \mathbf{A}_q(2,2)\mathbf{i}_s(k)\\ &\quad + \mathbf{B}_q(2,1)\mathbf{v}_s(k) + \mathbf{B}_q(2,2)\mathbf{i}_e(k)\end{aligned} \tag{7.31}$$

At this point, the method can use this model to predict the value of $\mathbf{i}_s$ depending on $\mathbf{i}_i$. Then (7.5) must be used to calculate $\mathbf{i}_i$ for each switching state.

7.4.1.4 The Instantaneous Reactive Power

The behavior of the MC can be improved by considering the instantaneous reactive power Q in the three-phase grid.

This reactive power is calculated by the following equation [6]:

$$Q = Im\{\mathbf{v}_s(t) \cdot \bar{\mathbf{i}}_s(t)\} \tag{7.32}$$

where $Im\{\}$ corresponds to the imaginary part of the product of the vectors and $\bar{\mathbf{i}}_s(t)$ is the complex conjugate of $\mathbf{i}_s(t)$.

The instantaneous reactive power can be predicted by using

$$\begin{aligned}Q(k+1) &= Im\{\mathbf{v}_s(k+1) \cdot \bar{\mathbf{i}}_s(k+1)\}\\ &= v_{s\beta}(k+1)i_{s\alpha}(k+1) - v_{s\alpha}(k+1)i_{s\beta}(k+1)\end{aligned} \tag{7.33}$$

where the subscripts α and β represent the real and imaginary components of the associated vector. The predicted value for the input current $\mathbf{i}_s(k+1)$ is obtained from (7.31). Line voltages are low-frequency signals and it can be considered that $\mathbf{v}_s(k+1) \approx \mathbf{v}_s(k)$.

7.4.2 Output Current Control

A block diagram of the predictive current control scheme for the MC is shown in Figure 7.5 considering an induction machine (IM) as the load. The load model and the filter model are used for calculating predictions of the future values of the output current for the 27 switching states of the MC. Based on these predictions, a cost function is used for selection of the optimal switching state to be applied in the converter.

The cost function g represents the evaluation criterion by which the control method determines the optimum switching state to be applied during the next sampling time. The optimization process is performed by evaluating g for each valid switching state. But prior to determining the equation that will represent the cost function, it is necessary to define the objectives that the converter must achieve. The MC must feed the load with currents close to the reference value and, at the same time, allow control of the input current in order to get low harmonic distortion and regulated power factor (PF). Several other objectives can be included in the cost function due to the versatility of the presented approach, opening up a wide range of possibilities for further research.

One of the objectives reflected in the cost function is the tracking of the reference current to the load. Switching states that generate closer values of the output current to the reference should be preferred. That goal is achieved by assigning cost or penalizing

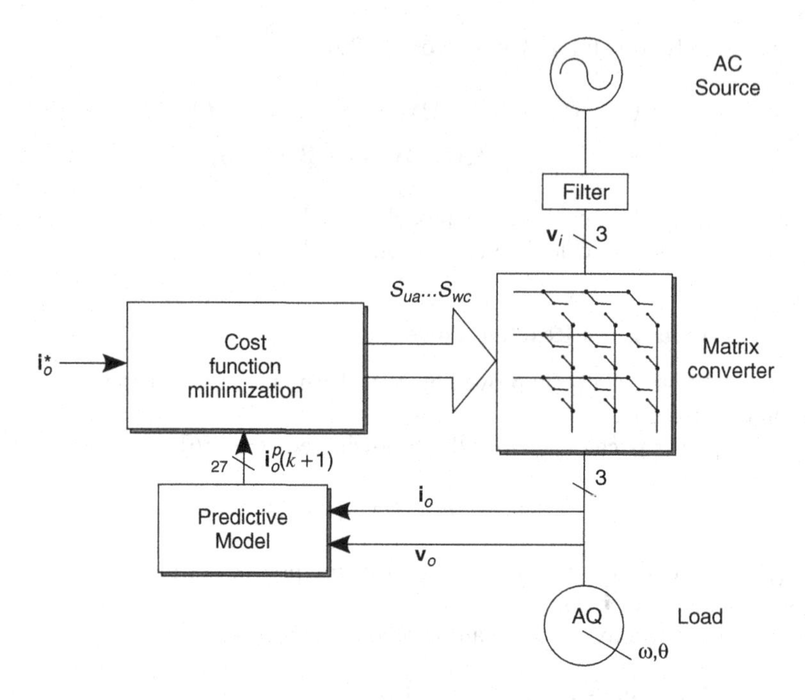

Figure 7.5 Block diagram of the predictive current control strategy

differences from the reference value, as expressed in the following cost function:

$$g_1 = |i_{o\alpha}^*(k+1) - i_{o\alpha}^{\mathbf{p}}(k+1)| + |i_{o\beta}^*(k+1) - i_{o\beta}^{\mathbf{p}}(k+1)| \tag{7.34}$$

where currents $i_{o\alpha}^*$ and $i_{o\beta}^*$ are the real and imaginary parts of the reference current vector $\mathbf{i}^*$. Currents $i_{o\alpha}^p$ and $i_{o\beta}^p$ are the real and imaginary parts of the predicted current vector $\mathbf{i}^p(k+1)$ calculated using (7.24) for a given switching state.

The behavior of the predictive control scheme for the MC is shown next for an 11 kW induction machine fed by an 18 kVA MC.

The predictive control scheme implemented without control of reactive power presents good reference tracking for the output currents, as shown in Figure 7.6. However, the input currents are very distorted and the input power factor is not controlled.

One of the advantages of the MC topology is the capacity for regeneration, that is, power flowing from the load to the grid. The operation of the MC in regeneration conditions is shown in Figure 7.7 for current control without regulation of the input power factor. It can be observed that the output current control maintains its good performance while the input currents are distorted.

7.4.3 Output Current Control with Minimization of the Input Reactive Power

The MC can control, in addition to the output currents, the phase of the input current from the mains. The amplitude is determined by the active power flow, since the MC does

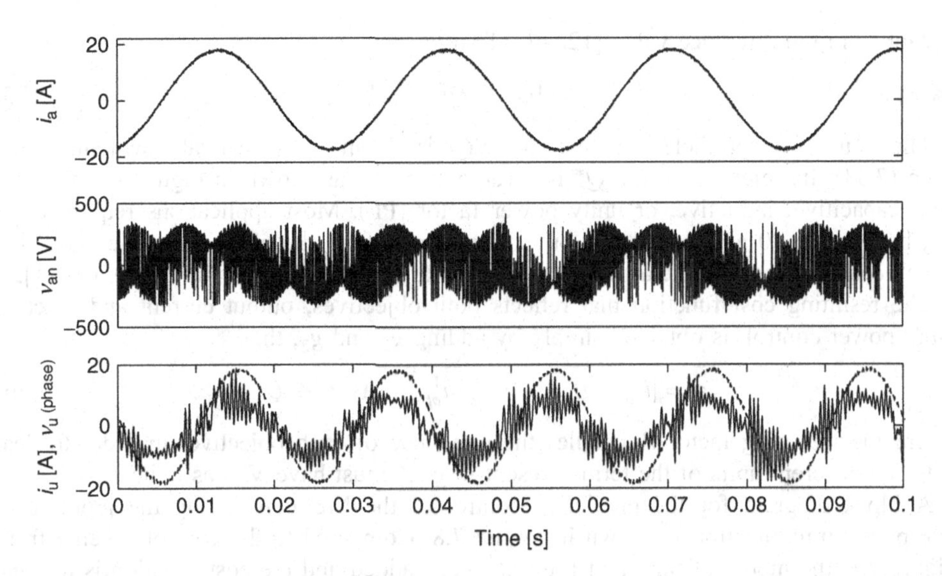

Figure 7.6 Steady state operation of the load current control without control of the input reactive power (Vargas *et al.*, 2008 © IEEE)

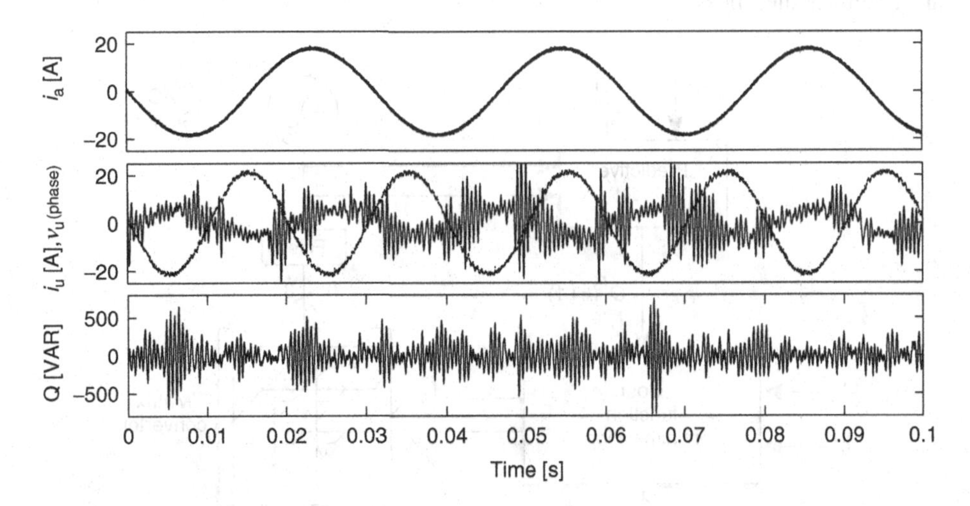

Figure 7.7 Steady state in regeneration without control of the input reactive power (Vargas *et al.*, 2008 © IEEE)

not store energy. Most modulation methods reach the objective of working with unity power factor by means of relatively complex strategies [7, 8]. In order to work with inductive or capacitive power factor, the complexity increases considerably [9–11]. With the predictive control approach, the objective of controlling the reactive power Q can be easily achieved simply by penalizing switching states that produce predictions of Q

distant from the reference value [12, 13]. That is

$$g_2 = |Q^* - Q^{\mathrm{p}}(k+1)| \tag{7.35}$$

The value of the predicted reactive power Q^{p} is obtained for each valid switching state from (7.33). Its reference value Q^* is given externally, as shown in Figure 7.5, to work with capacitive, inductive, or unity power factor (PF). Most applications require unity PF, hence $Q^* = 0$ and $g_2 = |Q^{\mathrm{p}}|$, but this method offers the alternative to control that variable with a very simple approach compared to other modulation strategies [9–11].

The resulting cost function that reflects both objectives, output current and reactive input power control, is obtained simply by adding g_1 and g_2, thus

$$g = |i^*_{o\alpha} - i^{\mathrm{p}}_{o\alpha}| + |i^*_{o\beta} - i^{\mathrm{p}}_{o\beta}| + A|Q^* - Q^{\mathrm{p}}| \tag{7.36}$$

where the weighing factor A handles the relevance of each objective. In order to deal with the different units of the terms present in g, A must have V^{-1} as unit.

A block diagram for the predictive control of the MC considering the input reactive power minimization is shown in Figure 7.8. Compared to the control scheme from Figure 7.5, the model of the input filter has been added and the cost function is the one presented in (7.36).

A higher value of A implies a higher relevance for controlling the PF or reactive input power over the output current reference tracking. The criteria for selecting the value of A are briefly treated next.

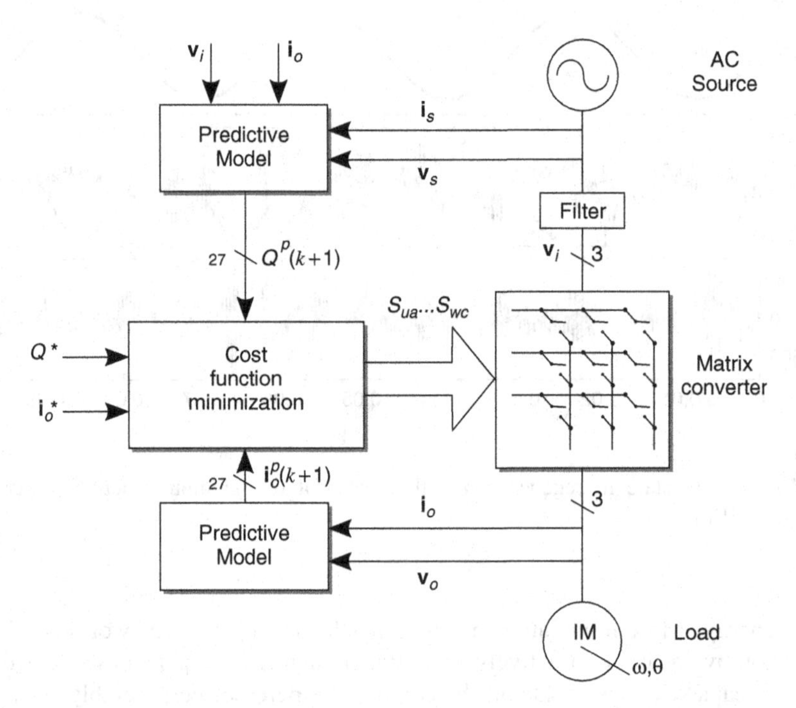

Figure 7.8 Block diagram of the predictive current control strategy with minimization of the input reactive power

7.4.3.1 Selection of Weighting Factor *A*

Weighting factor *A* is the only parameter from the predictive current controller to be adjusted. The adjustment of this kind of parameter is still an open topic for research. It is possible to find optimal values in cases where the system presents no constraints and under specific structures of the cost function [14]. For systems with a finite number of control actions, finite input sets, or state alphabet [15], one method to adjust the parameter is to simply evaluate the performance of specific system variables by simulations to determine the best value.

For the presented MC-based system, key variables for evaluating the behavior of the predictive current control method are the total harmonic distortion (THD) of the output and input or mains current. In order to perform the evaluation, an exhaustive search was carried out based on 400 simulations, each with an equidistant value of A within the range $0–0.007\,\mathrm{V}^{-1}$. As mentioned, the variables to observe and evaluate in order to select the weighing factor are the input and output current THD. The result of this procedure is shown in Figure 7.9.

As expected, the input current's THD drastically decreases as A increases, reaching a value close to 5% near $A = 0.002\ \mathrm{V}^{-1}$. No further reduction is significant after that value of A. On the other hand, as a trade-off, the output current's THD increases as more importance is placed on the reactive power in the cost function as A increases. Although the output current's THD is still low within the evaluated range–values from 0.09% to 0.14% are not considered to be high distortion–the ideal situation is to achieve low distortion on both currents. For that reason, the weighing factor A was set at $A = 0.0045\,\mathrm{V}^{-1}$, to select a value far from the region where the input current's THD drastically increases (under $A = 0.002\,\mathrm{V}^{-1}$) and not higher, in order to ensure low THD on the output current, according to the results shown in Figure 7.9.

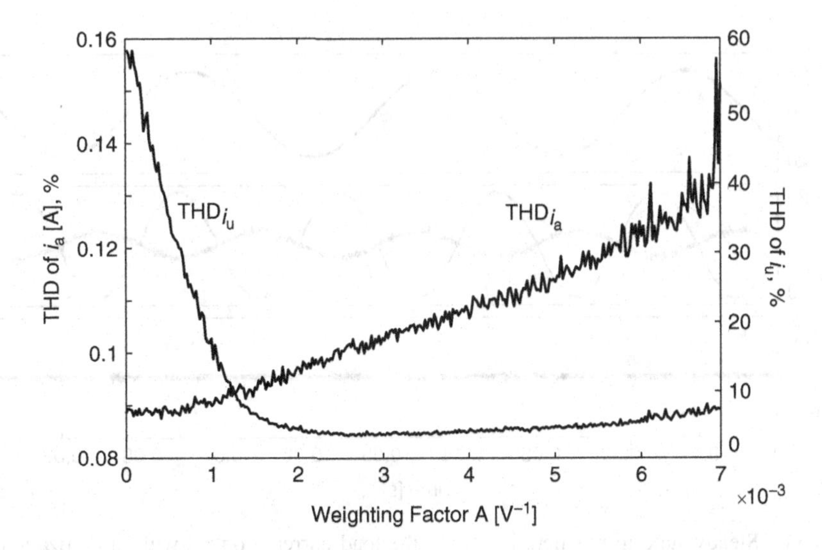

Figure 7.9 Exhaustive search to evaluate and select parameter *A* (Vargas *et al.*, 2008 © IEEE)

In order to obtain unity PF a reactive power reference $Q^* = 0$ is set. Results for these conditions are shown in Figure 7.10. It can be observed that the good quality of the output current control is maintained, compared to the results of Figure 7.6, while the waveform of the input currents is greatly improved. The input currents are sinusoidal with low distortion and in phase with the grid voltages. Control of the input PF is achieved even during regenerative operation of the converter, as shown in Figure 7.11.

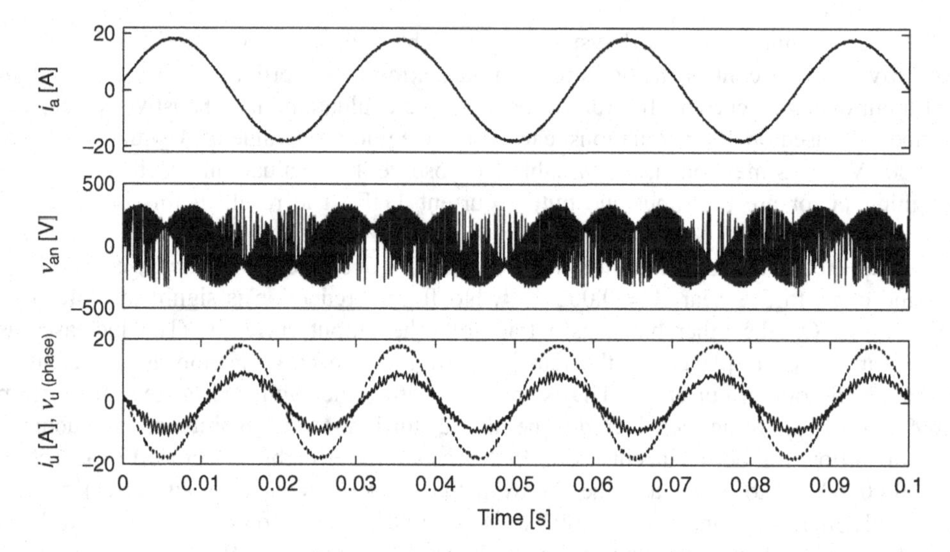

Figure 7.10 Steady state operation of the load current control with minimization of the input reactive power ($A = 0.0045$) (Vargas *et al.*, 2008 © IEEE)

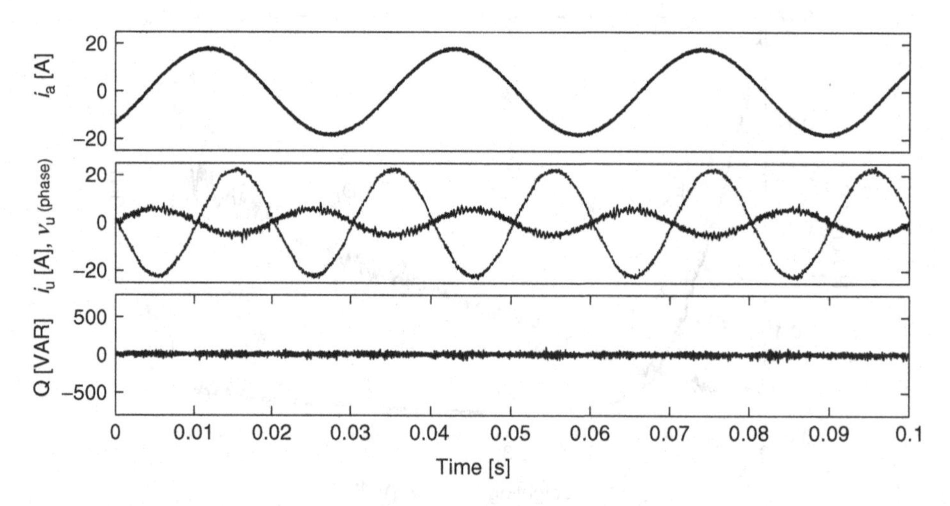

Figure 7.11 Steady state in regeneration using the load current control with minimization of the input reactive power ($A = 0.0045$) (Vargas *et al.*, 2008 © IEEE)

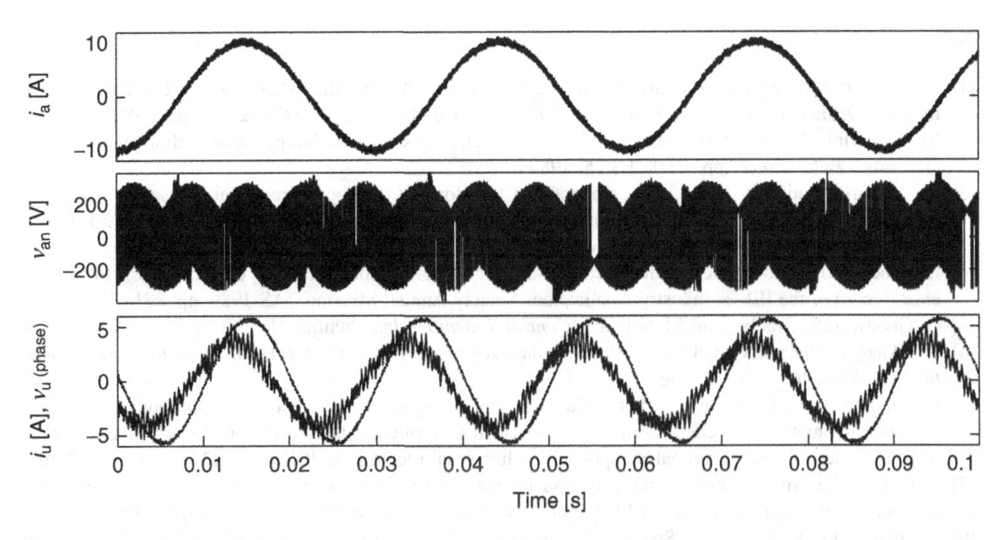

Figure 7.12 Performance of the drive with capacitive PF ($Q^* = 900$ VAR), $A = 0.0045$ (Vargas *et al.*, 2008 © IEEE)

7.4.4 Input Reactive Power Control

The capability of this control method to regulate the input PF is shown in Figure 7.12 for a reactive power reference of $Q^* = 900$ VAR. The parameter A is set to 0.0045.

Results for $Q^* = 0$ VAR (unity PF) can be seen in Figure 7.10. No difference is observed in terms of the output behavior of the converter. The main difference, as expected, is in the behavior of the input current, presenting a capacitive PF of 0.81. Consequently, it is possible to control the input PF by changing the value of Q^*.

7.5 Summary

This chapter presents a predictive control scheme for a matrix converter that effectively controls the output currents and the reactive input power. The strategy presented allows the input power factor to be regulated by means of a simple and straightforward technique, controlling the phase of the input current in such a way that the converter can work with capacitive, unity, or inductive power factor, according to the requirements of the application.

This method can be easily implemented by taking advantage of the present technologies available in digital signal processors. The high sampling frequency required should not be a problem nowadays and even less in years to come. This control strategy uses, in a convenient way, the discrete nature of power converters and the microprocessors used in their control.

The predictive control scheme is simpler than the classical control scheme for the matrix converter.

References

[1] P. Wheeler, J. Rodríguez, J. Clare, L. Empringham, and A. Weinstein, "Matrix converters: a technology review," *IEEE Transactions on Industrial Electronics*, vol. 49, no. 2, pp. 276–288, April 2002.

[2] M. Venturini, "A new sine wave in, sine wave out, conversion technique which eliminates reactive elements," *Powercon 7*, pp. E3/1–E3/15, 1980.

[3] M. Venturini and A. Alesina, "The generalized transformer: a new bidirectional sinusoidal waveform frequency converter with continuously adjustable input power factor," in Proceedings of the IEEE Power Electronics Specialists Conference, PESC'80, pp. 242–252, 1980.

[4] L. Hubert and D. Borojevic, "Space vector modulator for forced commutated cycloconverters," in Conference Record of the IEEE–Industry Applications Society Annual Meeting, IAS-1989, pp. 871–876, 1989.

[5] G. Goodwin, S. Graebe, and M. Salgado, *Control System Design* Prentice Hall, 2001.

[6] H. Akagi, E. Watanabe, and M. Aredes, *Instantaneous Power Theory and Applications to Power Conditioning*. Wiley-Interscience, 2007.

[7] P. Gamboa, S. Ferreira Pinto, J. Fernando Silva, and E. Margato, "Predictive optimal control of unity-power-factor matrix converters used in field oriented controlled induction motor drives," in Conference Record of the IEEE International Symposium on Industrial Electronics, ISIE, pp. 1334–1339, June 2007.

[8] S. Pinto and J. Silva, "Direct control method for matrix converters with input power factor regulation," in Conference Record of PESC, IEEE PE Society Annual Meeting, pp. 2366–2372, June 2004.

[9] L. Huber and D. Borojević, "Space vector modulated three-phase to three-phase matrix converter with input power factor corrections," *IEEE Transaction on Industry Applications*, vol. 31, no. 6, pp. 1234–1246, November/December 1995.

[10] D. Casadei, G. Serra, A. Tani, and L. Zarri, "Matrix converter modulation strategies: a new general approach based on space-vector representation of the switch state," *IEEE Transactions on Industrial Electronics*, vol. 49, no. 2, pp. 370–381, April 2002.

[11] Y.-D. Yoon and S.-K. Sul, "Carrier-based modulation method for matrix converter with input power factor control and under unbalanced input voltage conditions," in Conference Record of the IEEE Applied Power Electronics Conference, APEC 2007.

[12] R. Vargas, J. Rodríguez, U. Ammann, and P. Wheeler, "Predictive current control of an induction machine fed by a matrix converter with reactive power control," *IEEE Transactions on Industrial Electronics*, vol. 55, no. 12, pp. 4362–4371, December 2008.

[13] S. Muller, U. Ammann, and S. Rees, "New time-discrete modulation scheme for matrix converters," *IEEE Transactions on Industrial Electronics*, vol. 52, no. 6, pp. 1607–1615, December 2005.

[14] E. F. Camacho and C. Bordons, *Model Predictive Control* Springer Verlag, 1999.

[15] D. Quevedo, J. D. Dona, and G. Goodwin, "On the dynamics of receding horizon linear quadratic finite alphabet control loops," IEEE Conference on Decision and Control, pp. 2929–2934, December 2002.

Part Three

Model Predictive Control Applied to Motor Drives

8

Predictive Control of Induction Machines

8.1 Introduction

Over recent decades, control of electrical drives has been widely studied. Linear methods like PI controllers using PWM and non linear methods such as hysteresis control have been fully documented in the literature and dominate high-performance industrial applications [1, 2]. The most widely used linear strategy in high performance electrical drives is field-oriented control (FOC) [3–6], in which a decoupled torque and flux control is performed by considering an appropriate coordinate frame. A non linear hysteresis-based strategy such as direct torque control (DTC) [7] appears to be a solution for high performance applications.

At the end of the 1970s, model predictive control (MPC) was developed in the petro-chemical industry [8–10]. The term MPC does not imply a specific control strategy, but covers an ample variety of control techniques that make explicit use of a mathematical model of the process and minimization of an objective function [11] to obtain the optimal control signals. The slow dynamics of chemical processes allow long sample periods, providing enough time to solve the online optimization problem.

Due to the rapid development of microprocessors, the idea of having only a centralized controller, without a cascade control structure, was considered to improve the dynamic behavior. Furthermore, the increasing number of drive applications, in which fast dynamic response, low parameter sensitivity, and algorithm simplicity are required, has motivated the development of new control strategies capable of improving the performance. The first ideas on MPC applied to power converters and drives originated in the 1980s [12, 13].

The concept of MPC is based on the calculation of the future behavior of the system, in order to use this information to calculate optimal values for the actuating variables. Execution of the predictive algorithm can be divided into three main steps: *estimation* of the variables that cannot be measured, *prediction* of the future behavior of the system, and *optimization* of outputs, according to a previously designed control law.

Predictive Control of Power Converters and Electrical Drives, First Edition. Jose Rodriguez and Patricio Cortes.

For motor drive applications, the measured variables $\mathbf{i}_s$, ω, and a mathematical model of the machine are used to estimate the variables that cannot be measured, such as the rotor and stator flux $\boldsymbol{\psi}_r$, $\boldsymbol{\psi}_s$. Then, the same model is used to predict the future behavior of the variables for every control action. Finally, the voltage vector that produces the optimum reference tracking is selected as the switching state for the next sampling step. The model of the machine is the most important part of the controller, because both estimations and predictions depend on it.

Predictive control has many advantages that make it a real option if high dynamic control of electrical drives is required. The concept is easy to understand and implement, constraints and nonlinearities can be included, and multivariable cases can be considered. This control scheme requires lots of calculations compared to traditional strategies. Fortunately, the performance of current processors is sufficiently powerful to make this approach possible. The main difference between predictive control and traditional strategies is the precalculation of the system behavior, and its consideration in the control algorithm before the difference between the reference and the measured value occurs. The feed back PI- control loop corrects the control difference when it has already appeared.

This chapter presents two different approaches for the use of predictive control for induction machines. The first one is based on FOC and considers a predictive current control loop. The second approach is called predictive torque control (PTC) and uses a model of the system and an appropriate cost function to directly control torque and flux [14, 15]. In order to illustrate the flexibility of the predictive control schemes both approaches, FOC and PTC, are presented for a simple three-phase inverter and a more complex matrix converter.

8.2 Dynamic Model of an Induction Machine

A three-phase current system can be represented by a three-axis coordinate system, as shown in Figure 8.1a. Unfortunately, the three axes are linearly dependent, a fact that

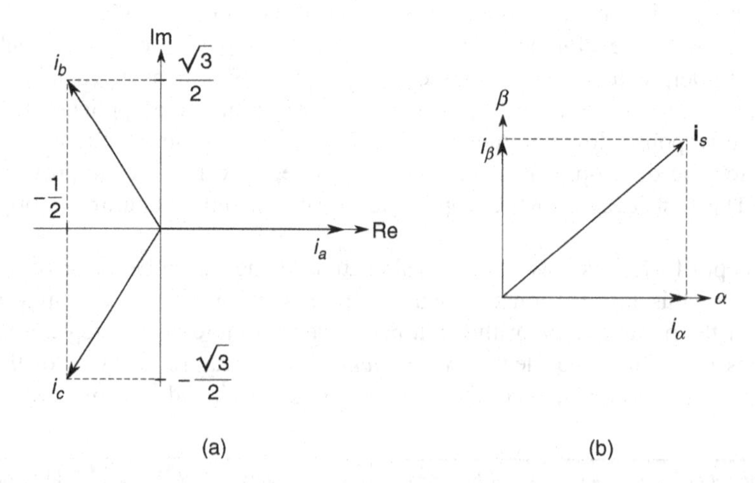

Figure 8.1 Coordinate transformation. (a) Currents expressed in a three-phase reference frame (a, b, c). (b) Currents expressed in a complex reference frame $\alpha\beta$

makes the mathematical description of a three-phase machine difficult. However, the linear dependence means that only two variables are necessary to describe three physical quantities. Hence, a complex, linearly independent coordinate system can be selected. Figure 8.1b shows the equivalent representation.

A three-phase stator currents system, with angular frequency ω_0, can be defined in a fixed three-phase coordinate frame:

$$i_a = I \cdot \sin(\omega_0 t) \tag{8.1}$$

$$i_b = I \cdot \sin\left(\omega_0 t + \frac{2\pi}{3}\right) \tag{8.2}$$

$$i_c = I \cdot \sin\left(\omega_0 t + \frac{4\pi}{3}\right) \tag{8.3}$$

For the stator currents, the transformation from a three- to a two-phase system is described by

$$\mathbf{i}_s = \frac{2}{3}(i_a + \mathbf{a} i_b + \mathbf{a}^2 i_c) \tag{8.4}$$

$$\mathbf{a} = e^{j2\pi/3} = \frac{-1}{2} + j\frac{\sqrt{3}}{2} \tag{8.5}$$

$$\mathbf{a}^2 = e^{j4\pi/3} = \frac{-1}{2} - j\frac{\sqrt{3}}{2} \tag{8.6}$$

The same coordinate transformation shown above is used for the electromagnetic variables. Thus, the equations of an induction machine [16] can be represented in any arbitrary reference frame rotating at an angular frequency ω_k. The variable ω denotes the rotor angular speed:

$$\mathbf{v}_s = R_s \mathbf{i}_s + \frac{d\boldsymbol{\psi}_s}{dt} + j\omega_k \boldsymbol{\psi}_s \tag{8.7}$$

$$0 = R_r \mathbf{i}_r + \frac{d\boldsymbol{\psi}_r}{dt} + j(\omega_k - \omega)\boldsymbol{\psi}_r \tag{8.8}$$

$$\boldsymbol{\psi}_s = L_s \mathbf{i}_s + L_m \mathbf{i}_r \tag{8.9}$$

$$\boldsymbol{\psi}_r = L_m \mathbf{i}_s + L_r \mathbf{i}_r \tag{8.10}$$

$$T = \frac{3}{2} p Re\{\bar{\boldsymbol{\psi}}_s \mathbf{i}_s\} = -\frac{3}{2} p Re\{\bar{\boldsymbol{\psi}}_r \mathbf{i}_r\} \tag{8.11}$$

where:

- L_s, L_r, and L_m are the stator, rotor, and magnetizing inductances, respectively.
- R_s and R_r are the stator and rotor resistances.
- $\mathbf{v}_s$ and $\mathbf{i}_s$ are the stator voltage and current vectors.
- $\mathbf{i}_r$ is the rotor current vector.
- $\boldsymbol{\psi}_s$ and $\boldsymbol{\psi}_r$ are the stator and rotor flux vectors.
- T and p are electromagnetic torque and number of pole pairs, respectively.
- $\bar{\boldsymbol{\psi}}$ is the complex conjugate value of $\boldsymbol{\psi}$.

In (8.8), the rotor vector voltage $\mathbf{v}_r$ is equal to zero because a squirrel-cage motor is considered. Hence, the rotor winding is short-circuited.

If the mechanical equation of the rotor is considered in (8.12), it is possible to see that the torque affects the ratio of change in the mechanical rotor speed ω_m:

$$J\frac{d\omega_m}{dt} = T - T_l \tag{8.12}$$

The coefficient J in (8.12) denotes the moment of inertia of the mechanical shaft, and T_l is the load torque connected to the machine; it corresponds to an external disturbance, which must be compensated by the control system. ω_m is the mechanical rotor speed, which is related to the electric rotor speed ω by the number of pole pairs p:

$$\omega = p\omega_m \tag{8.13}$$

In order to develop an appropriate control strategy, it is convenient to write the equations of the machine in terms of state variables. The stator current $\mathbf{i}_s$ and the rotor flux $\boldsymbol{\psi}_r$ vectors are selected as state variables. The stator current is especially selected because it is a variable that can be measured, and also undesired stator dynamics, like effects on the stator resistance, stator inductance, and back-emf, are avoided. Thus, according to [17] and [18], the equivalent equations of the stator and rotor dynamics of a squirrel-cage induction machine are obtained:

$$\mathbf{i}_s + \tau_\sigma\frac{d\mathbf{i}_s}{dt} = -j\omega_k\tau_\sigma\mathbf{i}_s + \frac{k_r}{R_\sigma}\left(\frac{1}{\tau_r} - j\omega\right)\boldsymbol{\psi}_r + \frac{\mathbf{v}_s}{R_\sigma} \tag{8.14}$$

$$\boldsymbol{\psi}_r + \tau_r\frac{d\boldsymbol{\psi}_r}{dt} = -j(\omega_k - \omega)\tau_r\boldsymbol{\psi}_r + L_m\mathbf{i}_s \tag{8.15}$$

where

$$\tau_s = \frac{L_s}{R_s}$$

$$\tau_r = \frac{L_r}{R_r}$$

$$\sigma = 1 - \frac{L_m^2}{L_s L_r}$$

$$k_r = \frac{L_m}{L_r}$$

$$k_s = \frac{L_m}{L_s}$$

$$R_\sigma = R_s + R_r k_r^2$$

$$\tau_\sigma = \frac{\sigma L_s}{R_\sigma}$$

These equations will be used for estimating the stator and rotor flux, and for calculating predictions for the stator currents, stator flux, and electrical torque using the appropriate discrete-time version of the equations.

8.3 Field Oriented Control of an Induction Machine Fed by a Matrix Converter Using Predictive Current Control

When the induction machine is fed by a matrix converter, the same predictive control strategy presented in Chapter 7 for this converter can be used as part of a FOC scheme. However, specific requirements for the matrix converter can be included in the cost function.

8.3.1 Control Scheme

The control strategy for the matrix converter counts with an inner control loop that performs predictive current control, and an outer loop that handles speed, flux, and torque control by means of FOC [19], which generates reference currents for the predictive current control. A block diagram of the entire control strategy is presented in Figure 8.2.

A PI controller receives the speed error and generates the reference torque. The reference amplitude of the stator flux and reference torque are used, by means of FOC, to generate the output reference current to the predictive current control segment of the control strategy, as shown in Figure 8.2. The predictive current control strategy replaces, in this approach, linear current controllers and the modulation techniques of classic methods [20, 21]. The predictive algorithm handles the objectives of controlling the output current according to the reference signal received from the previous stage and regulating the input power factor to maintain the reactive input current close to its reference. Typically, most applications require unity power factor, but the presented approach also allows for capacitive and inductive power factors.

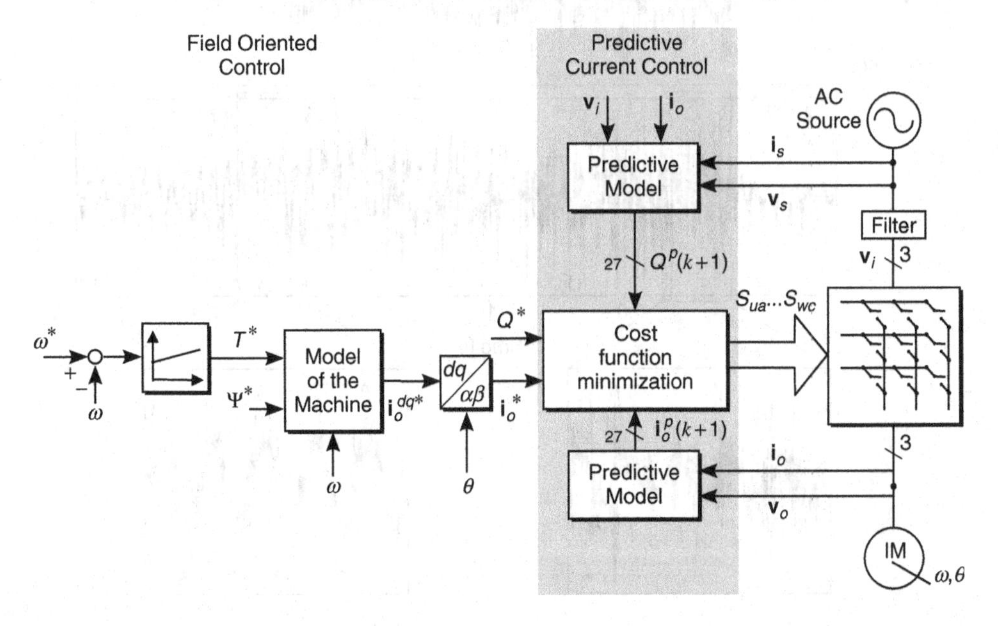

Figure 8.2 FOC of an induction machine fed by a matrix converter using predictive current control

On a matrix converter, there are 27 valid switching combinations [22]. The predictive current control method consists of choosing, at fixed sampling intervals, the best possible switching state of the converter, based on an evaluation criterion and predictions of the behavior of the system. For that purpose, the algorithm performs a cost function minimization by means of predictions of variables obtained from a model of the system. The nonlinear optimization problem is solved in real time by means of an exhaustive

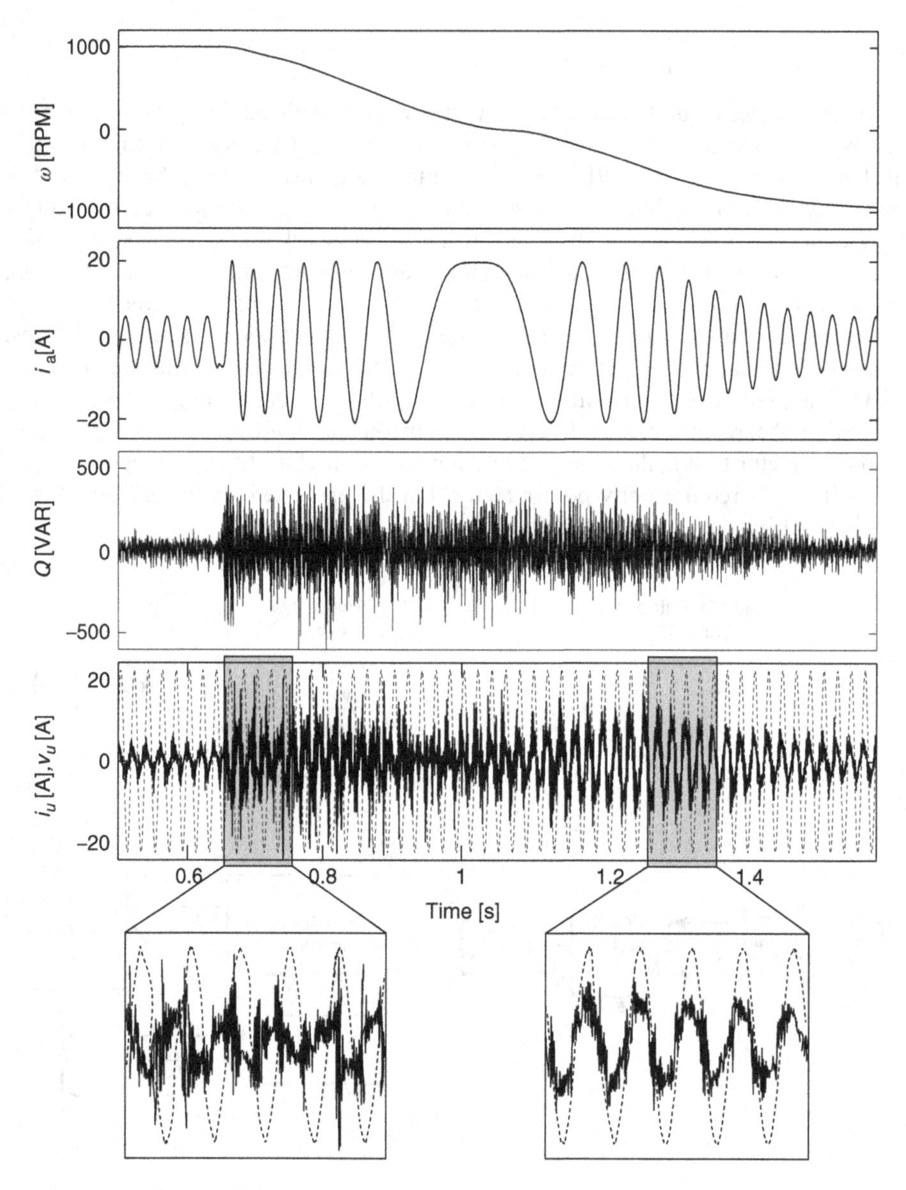

Figure 8.3 Speed reversal for the FOC strategy using a matrix converter without control of the input reactive power (Vargas *et al.*, 2008 © IEEE)

search process, that is, simply evaluating the cost function for each of the 27 valid switching states. The cost function is the evaluation criterion for the predictions and represents the desired behavior of the system.

For output current control and minimization of the input reactive power in the matrix converter, the cost function is defined as

$$g = |i_{o\alpha}^* - i_{o\alpha}^{\mathrm{p}}| + |i_{o\beta}^* - i_{o\beta}^{\mathrm{p}}| + A|Q^* - Q^{\mathrm{p}}| \tag{8.16}$$

As expressed in this cost function, the matrix converter requires control of the input side. Results for this predictive control strategy for an induction machine are shown in Figure 8.3 for speed reversal. A weighting factor value of $A = 0$ is considered for these results, that is, without control of the input side of the matrix converter. It can be observed that the performance of the output current control is very good, yielding a good operation of the drive. However, the input currents of the matrix converter are distorted.

The same test is performed for control of the input reactive power by setting the weighting factor to $A = 0.0045$. This value has been adjusted according to the criterion presented in Chapter 7. The behavior of the input currents is much improved, achieving near unity power factor, as shown in Figure 8.4.

8.4 Predictive Torque Control of an Induction Machine Fed by a Voltage Source Inverter

For an induction machine, it can be demonstrated that both the stator flux $\boldsymbol{\psi}_s$ and electromagnetic torque T can be modified by selecting a proper voltage vector sequence that modifies the magnitude of the stator flux and at the same time increases or decreases the angle between the rotor and stator flux. These ideas correspond to the basics of direct torque control.

In predictive torque control (PTC), the same principle is used, but in this scheme predictions for the future values of the stator flux and torque are calculated. Hence, the reference condition, which is implemented by a cost function, considers the future behavior of these variables. Predictions are calculated for every actuating possibility and the cost function selects the voltage vector that optimizes the reference tracking. A block diagram of PTC is shown in Figure 8.5.

The block concerning *estimation* is used to calculate the current values of the variables that cannot be measured, such as the rotor flux $\boldsymbol{\psi}_r$ and the stator flux $\boldsymbol{\psi}_s$. Then, the predictive model computes the future values of controlled variables at the instant $k + 1$, in this case the stator flux $\boldsymbol{\psi}_s(k + 1)$ and the electromagnetic torque $T(k + 1)$. These predictions are calculated for every actuating possibility given by the inverter topology. If a two-level inverter is considered, eight switching states and seven different voltage vectors can be generated. Finally, the block concerning *minimization* chooses the optimum switching state which minimizes the corresponding cost function. This function contains the control law in order to achieve an appropriate torque and flux regulation.

In PTC, estimations of the stator flux $\boldsymbol{\psi}_s$ and the rotor flux $\boldsymbol{\psi}_r$, at the present sampling step, are required.

The stator flux estimation is based on the stator voltage equation:

$$\mathbf{v}_s = R_s \mathbf{i}_s + \frac{d\boldsymbol{\psi}_s}{dt} \tag{8.17}$$

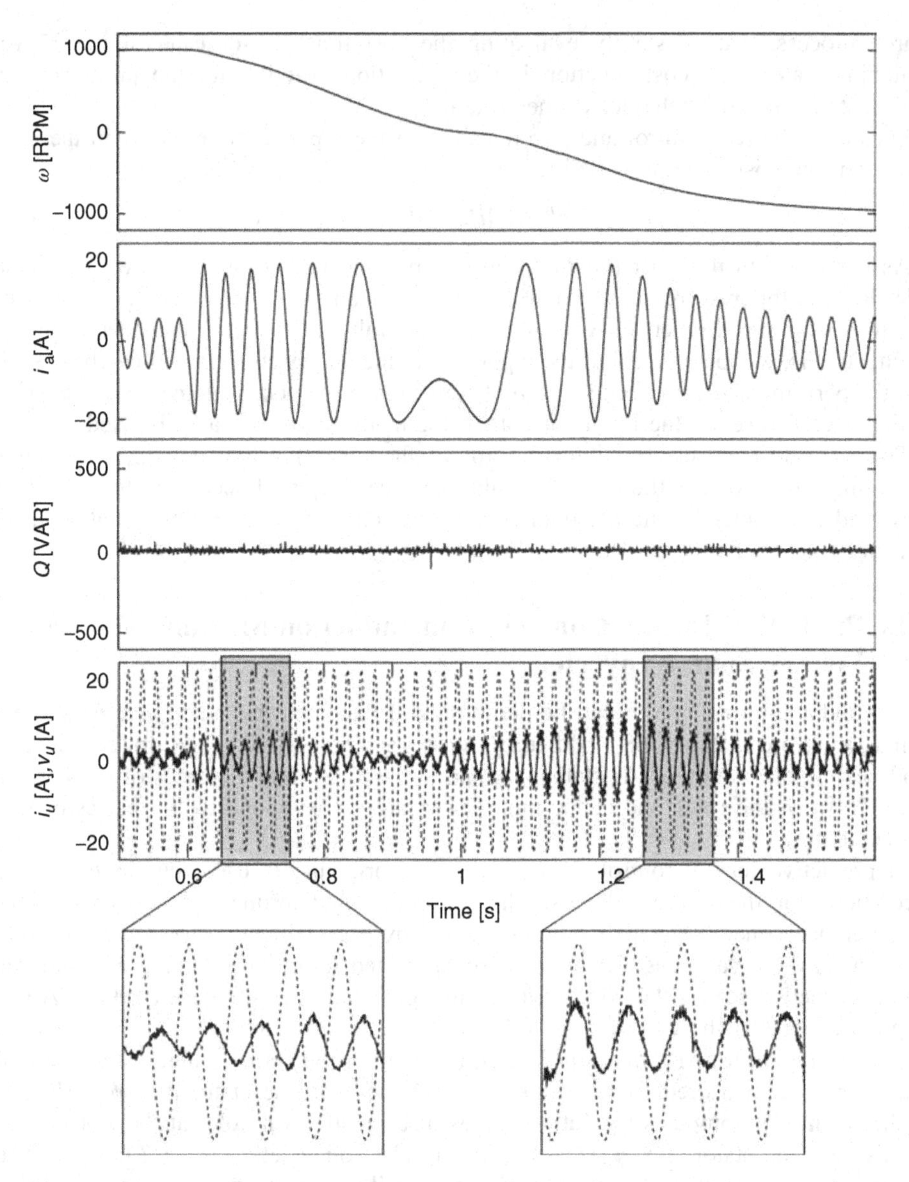

Figure 8.4 Speed reversal for the FOC strategy using a matrix converter with control of the input reactive power (Vargas *et al.*, 2008 © IEEE)

Using the Euler formula to discretize (8.17), the stator flux estimation is obtained:

$$\hat{\boldsymbol{\psi}}_s(k) = \hat{\boldsymbol{\psi}}_s(k-1) + T_s \mathbf{v}_s(k) - R_s T_s \mathbf{i}_s(k) \tag{8.18}$$

The rotor flux estimation $\hat{\boldsymbol{\psi}}_r(k)$ is obtained from the flux linkage equations, by replacing the rotor current $\mathbf{i}_r$ obtained from (8.19) in (8.20):

$$\boldsymbol{\psi}_r = \mathbf{i}_s L_m + \mathbf{i}_r L_r \tag{8.19}$$

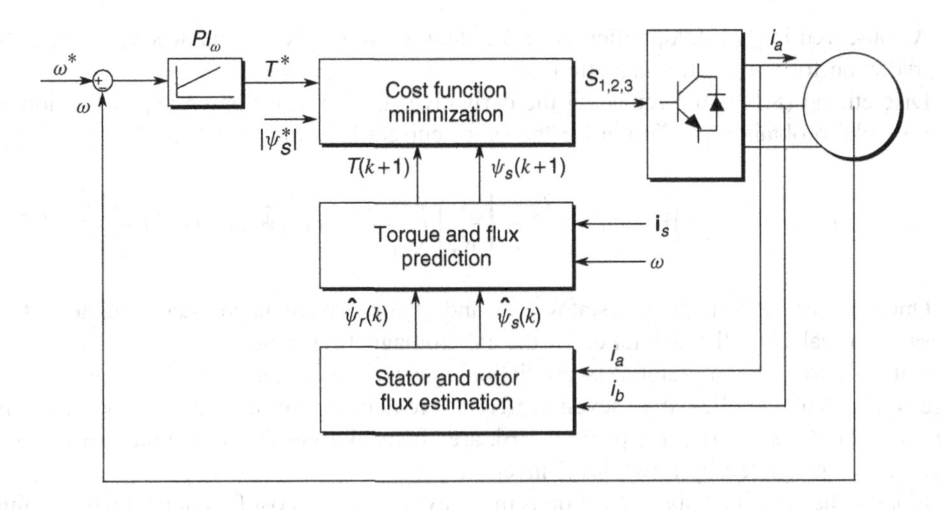

Figure 8.5 PTC scheme

$$\boldsymbol{\psi}_s = \mathbf{i}_s L_s + \mathbf{i}_r L_m \tag{8.20}$$

$$\hat{\boldsymbol{\psi}}_r = \frac{L_r}{L_m}\hat{\boldsymbol{\psi}}_s + \mathbf{i}_s\left(L_m - \frac{L_r L_s}{L_m}\right) \tag{8.21}$$

Thus, by discretizing (8.21) and replacing the current estimation for the stator flux $\hat{\boldsymbol{\psi}}_s(k)$, the rotor flux estimation $\hat{\boldsymbol{\psi}}_r(k)$ is obtained:

$$\hat{\boldsymbol{\psi}}_r(k) = \frac{L_r}{L_m}\hat{\boldsymbol{\psi}}_s(k) + \mathbf{i}_s(k)\left(L_m - \frac{L_r L_s}{L_m}\right) \tag{8.22}$$

After the rotor and stator flux estimations have been obtained, it is necessary to compute the predictions for the controlled variables. In the case of PTC, the electromagnetic torque T and the stator flux $\boldsymbol{\psi}_s$ are predicted for the next sampling instant $k+1$.

For the stator flux prediction $\boldsymbol{\psi}_s^p(k+1)$, the same stator voltage equation used for its estimation is considered. By approximating the stator flux derivative, the prediction for the stator flux is obtained:

$$\boldsymbol{\psi}_s^p(k+1) = \hat{\boldsymbol{\psi}}_s(k) + T_s\mathbf{v}_s(k) - R_s T_s \mathbf{i}_s(k) \tag{8.23}$$

The torque prediction depends directly on the stator flux and current according to

$$T = \frac{3}{2}pIm\{\bar{\boldsymbol{\psi}}_s \mathbf{i}_s\} \tag{8.24}$$

Thus, by considering the predicted values of the stator flux and stator current, the torque prediction is obtained:

$$T^p(k+1) = \frac{3}{2}pIm\{\bar{\boldsymbol{\psi}}_s^p(k+1)\mathbf{i}_s^p(k+1)\} \tag{8.25}$$

As observed in (8.25), a prediction of the stator current $\mathbf{i}_s^p(k+1)$ is needed to compute a prediction for the electromagnetic torque.

Discretizing (8.14), and replacing the derivatives by the Euler-based approximation, it is possible to obtain a prediction for the stator current $\mathbf{i}_s$ at the time $k+1$:

$$\mathbf{i}_s^p(k+1) = \left(1 + \frac{T_s}{\tau_\sigma}\right)\mathbf{i}_s(k) + \frac{T_s}{\tau_\sigma + T_s}\left\{\frac{1}{R_\sigma}\left[\left(\frac{k_r}{\tau_r} - k_r j\omega\right)\hat{\boldsymbol{\psi}}_r(k) + \mathbf{v}_s(k)\right]\right\} \tag{8.26}$$

Once the predictions of the stator flux and stator current have been obtained, it is possible to calculate the prediction of the electromagnetic torque.

Both the torque and stator flux predictions are written in terms of the inverter voltage $\mathbf{v}_s(k)$. This implies that seven different predictions for the torque and the flux $(T^p(k+1), \boldsymbol{\psi}_s^p(k+1))_h$, $h \in [0, 1, \ldots, 6]$, are obtained according to the number of voltage vectors generated by a two-level inverter.

Finally, the switching state selection is made by means of a cost function which contains the control law. Basically, it corresponds to a comparison between the torque and flux references to their predicted values. The cost function is evaluated for every prediction and the one that produces the lowest value is selected. Thus, the firing pulses of the inverter are generated.

The cost function has the following structure:

$$g_h = |T^* - T^p(k+1)_h| + \lambda_\psi |\boldsymbol{\psi}_s^* - \boldsymbol{\psi}_s^p(k+1)_h| \tag{8.27}$$

The term λ_ψ denotes the weighting factor, which increases or decreases the relative importance of the torque versus the flux control. This is the only parameter to be adjusted in PTC. If the same weight were assigned for both control variables, these factors would correspond to the ratio between the magnitudes of the nominal torque T_n and stator flux $|\boldsymbol{\psi}_{sn}|$:

$$\lambda_\psi = \frac{T_n}{|\boldsymbol{\psi}_{sn}|} \tag{8.28}$$

A dynamic result can be seen in Figure 8.6. The maneuver corresponds to controlled starting, a speed reversal, and a load torque response. It is possible to see that PTC achieves an excellent dynamic performance with low distortion of the stator currents and torque.

A torque step response is shown in Figure 8.7a. It is important to point out the fast dynamic behavior of the strategy. This can be explained by the fact that PTC is a direct strategy that does not require an inner PI control loop for the stator currents and modulators. Hence, there is no bandwidth limitation for the electromagnetic torque dynamics.

The behavior of the stator currents at steady state operation is presented in Figure 8.7b. In this condition the machine is operating at the rated speed (2860 RPM) at an equivalent load of 50% of the nominal torque. Note that low harmonic distortion can be achieved. The resulting THD is equal to 4%.

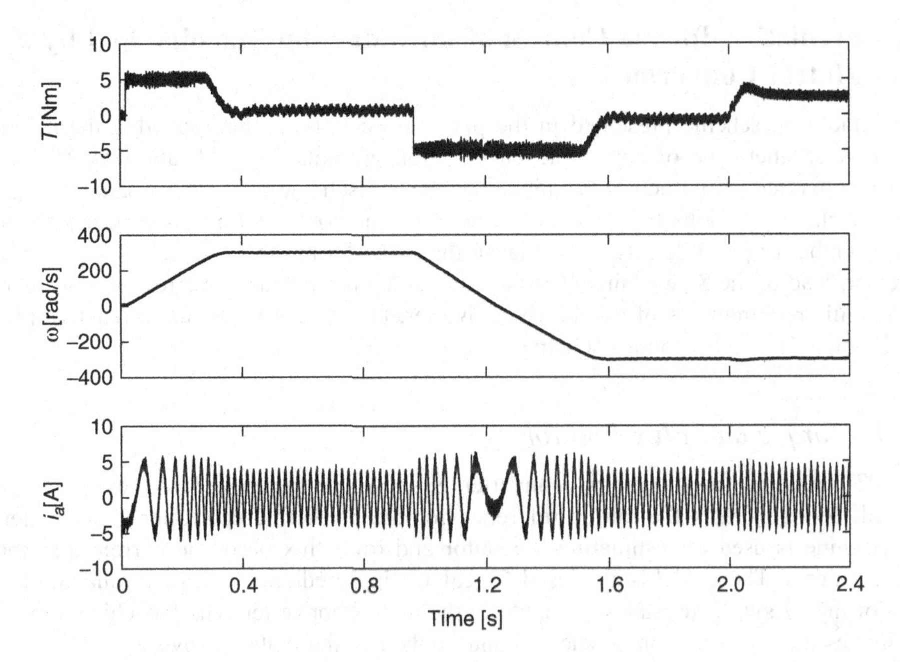

Figure 8.6 Starting, speed reversal, and load torque using PTC

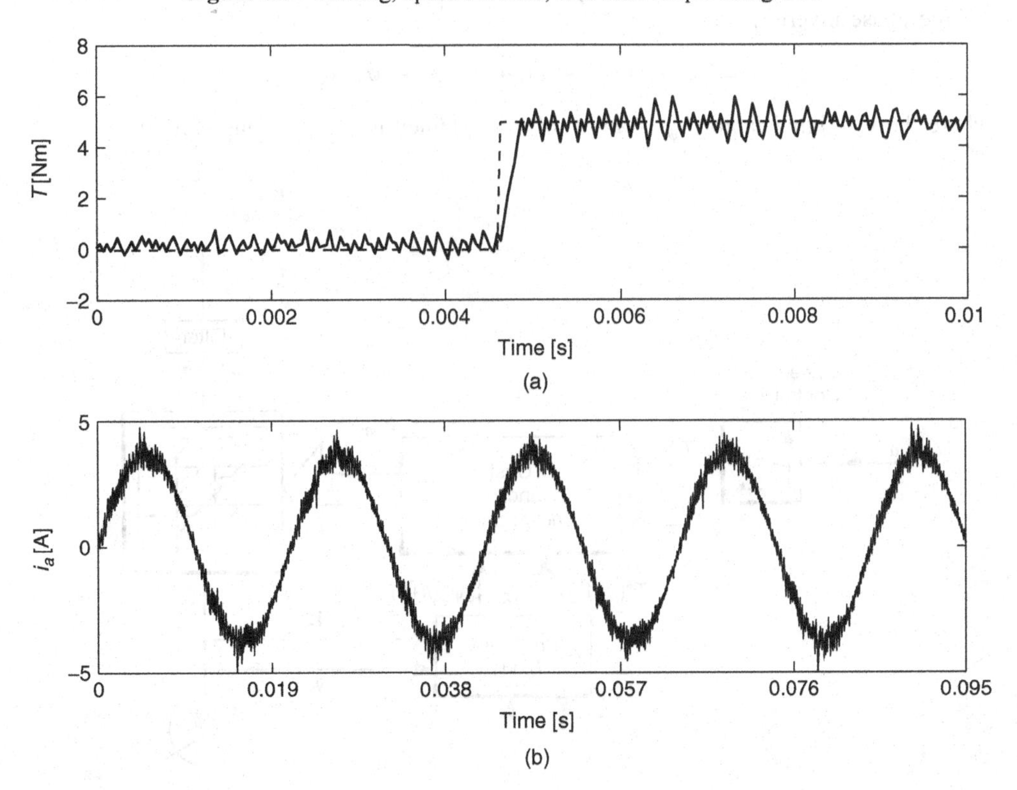

Figure 8.7 Torque step response and steady state behavior of the stator current. (a) Torque step response. (b) Stator current

8.5 Predictive Torque Control of an Induction Machine Fed by a Matrix Converter

The same PTC scheme presented in the previous section can be applied if the machine is fed by another type of converter. This section presents the application of PTC for a matrix converter-fed induction machine. The control scheme remains the same, using the same machine equations for predictions and the same cost function. The main difference is the number of possible actuations that in the case of a matrix converter is 27 switching states, instead of the 8 switching states available in a three-phase inverter. By considering the specific requirements of the matrix converter, the control of the input reactive power can be included in the control scheme.

8.5.1 Torque and Flux Control

The PTC scheme for the matrix converter is shown in Figure 8.8. An external speed control loop generates the torque reference for the predictive controller. The model of the machine is used for estimating the stator and rotor flux based on current and speed measurements. The model is also used for calculating predictions of the torque and stator flux for the 27 switching states available in the matrix converter. The switching state that minimizes the cost function is selected and applied in the matrix converter.

The cost function considers the torque and flux errors, and is same as the one used for a three-phase inverter:

$$g = |T^* - T^p(k+1)| + \lambda_\psi |\boldsymbol{\psi}_s^* - \boldsymbol{\psi}_s^p(k+1)| \tag{8.29}$$

where the weighting factor λ_ψ is adjusted as explained in the previous section.

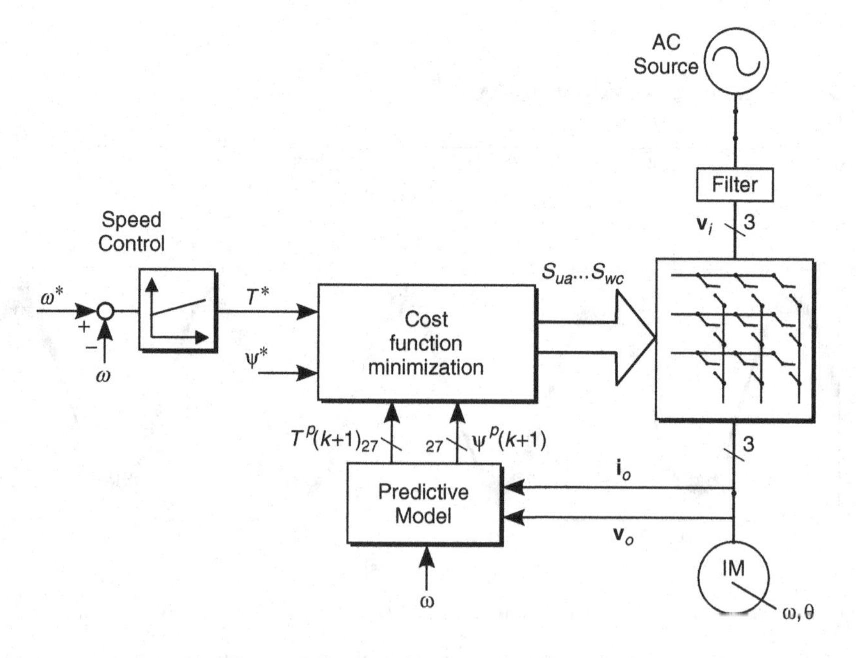

Figure 8.8 PTC for a matrix converter

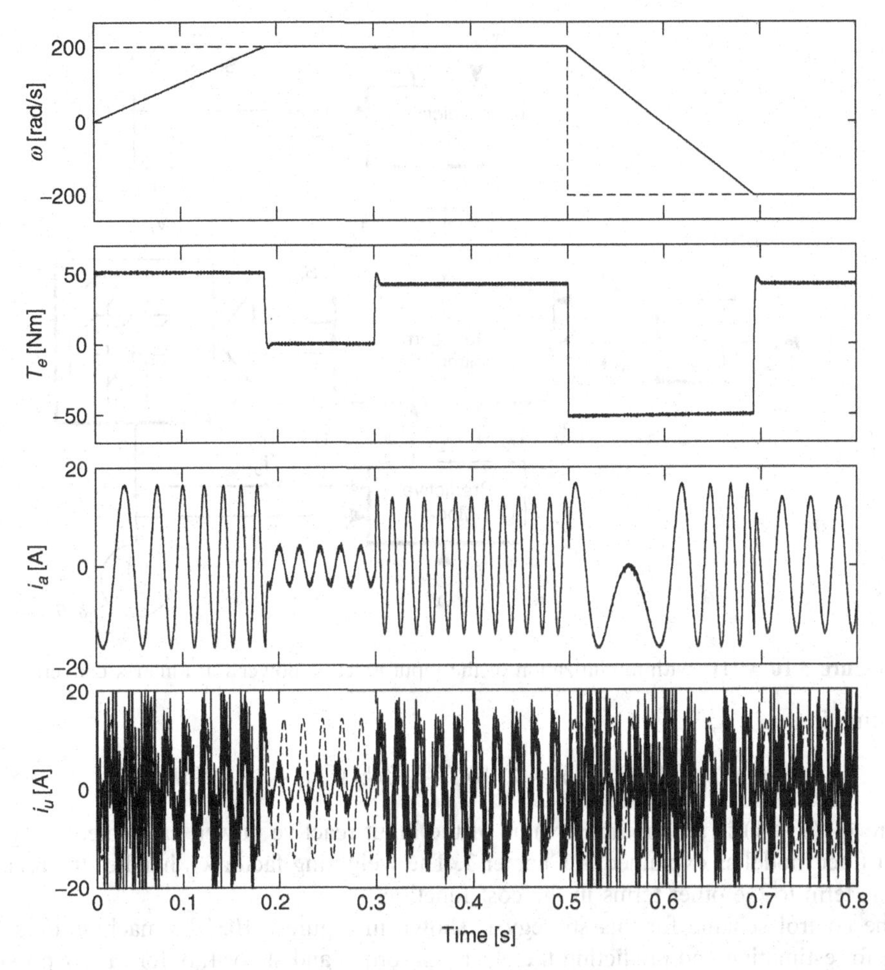

Figure 8.9 Speed reversal for the PTC without control of the input side of the matrix converter (Vargas *et al.*, 2008 © IEEE)

This control scheme allows fast control of the torque and flux of the induction machine, as shown in Figure 8.9 for a speed reversal. As the input reactive power is not controlled, it presents values different from zero, which is reflected in distorted input currents. This problem can be solved by including the reactive power minimization in the cost function as will be explained next.

8.5.2 Torque and Flux Control with Minimization of the Input Reactive Power

When PTC is implemented for a matrix converter, an additional requirement must be considered: the minimization of the input reactive power. This requirement can be easily considered in the control scheme by including an additional term in the cost function,

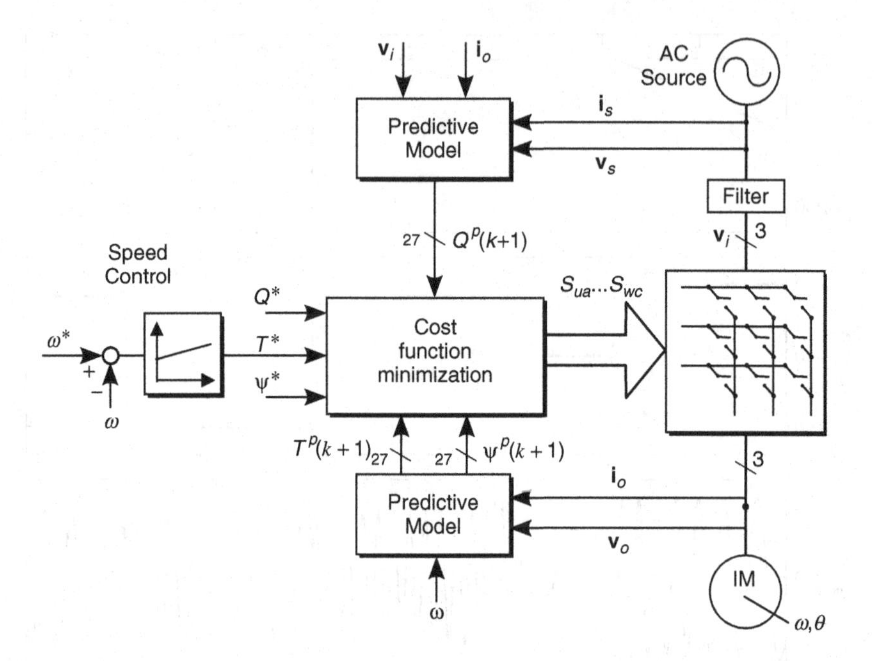

Figure 8.10 PTC with minimization of the input reactive power for a matrix converter

resulting in

$$g = |T^* - T^p(k+1)| + \lambda_\psi |\boldsymbol{\psi}_s^* - \boldsymbol{\psi}_s^p(k+1)| + \lambda_Q |Q^p(k+1)| \tag{8.30}$$

where $Q^p(k+1)$ is the predicted value of the input reactive power, calculated using the input filter model as explained in Chapter 7. The weighting factor λ_Q handles the relation of this term to the other terms in the cost function.

The control scheme for this strategy is shown in Figure 8.10. The machine model is used for estimating and predicting the electrical torque and stator flux for the 27 possible switching states. The filter model is used for calculating predictions for the input reactive power, for the 27 switching states. Then, the cost function is used for evaluation of the predictions, and the switching state that minimizes this function is selected and applied in the converter. The torque reference is generated by an external speed control loop.

The performance of the torque control is not affected by the modification of the cost function, as can be observed in Figure 8.11. However, the input currents are sinusoidal and in phase with the grid voltage, improving the behavior of the system with respect to the results shown in the previous section.

8.6 Summary

The application of MPC for the control of induction machines is presented in this chapter. Two types of control schemes are presented, based on the ideas of field-oriented control and direct torque control.

Field-oriented control is implemented by replacing the current controllers and modulator with a predictive current controller. This strategy is presented for a three-phase inverter

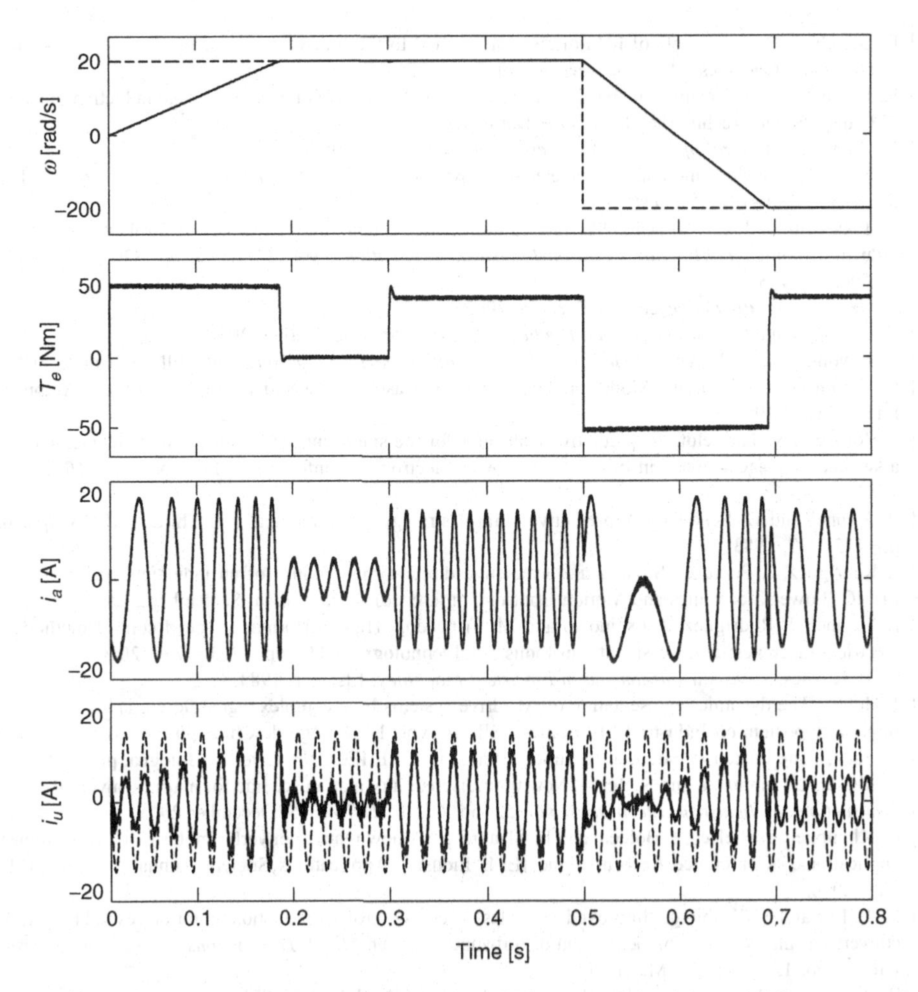

Figure 8.11 Speed reversal for the PTC with minimization of the input reactive power of the matrix converter (Vargas *et al.*, 2008 © IEEE)

and for a matrix converter. Good performance of the system is obtained due to the fast dynamic response of the predictive current control.

Predictive torque control achieves high-performance results which are totally comparable to the traditional approaches. This control strategy is presented for a three-phase inverter and for a matrix converter. As can be seen in this chapter, the control scheme remains the same and few changes are needed in order to implement it in a complex converter like the matrix converter.

References

[1] M. Kazmierkowski, R. Krishnan, and F. Blaabjerg, *Control in Power Electronics*. Academic Press, 2002.
[2] N. Mohan, T. Undeland, and W. Robbins, *Power Electronics*, 2nd ed. John Wiley & Sons, Inc. 1995.

[3] F. Blaschke, "The principle of field-orientation applied to the transvector closed-loop control system for rotating field machines," *Siemens Review*, vol. 34, 1972.

[4] K. Hasse, "On the dynamics of speed control of a static AC drive with a squirrel-cage induction machine," PhD dissertation. Technische Hochschule Darmstadt.

[5] W. Leonhard, *Control of Electrical Drives*. Springer Verlag, 1996.

[6] J. Holtz, "Pulsewidth modulation for electronic power conversion," *Proceedings of the IEEE*, vol. 82, no. 8, pp. 1194–1214, August 1994.

[7] I. Takahashi and T. Noguchi, "A new quick response and high efficiency control strategy for an induction motor," *IEEE Transaction on Industrial Applications*, vol. 22, no. 5, pp. 820–827, September/October 1986.

[8] J. Maciejowski, *Predictive Control* Pearson, 2002.

[9] E. Camacho and C. Bordons, *Model Predictive Control*, Springer Verlag, 2007.

[10] J. Rawlings and D. Mayne, *Model Predictive Control: Theory and Design*. Nob Hill Publishing 2009.

[11] M. Morari and J. H. Lee, "Model predictive control, past, present and future," Institut für Automatik, ETH, Zurich, 1998.

[12] J. Holtz and S. Stadtfeldt, "A predictive controller for the stator current vector of AC machines fed from a switched voltage source," in International Power Electronics Conference, IPEC, vol. 2, pp. 1665–1675, 1983.

[13] R. Kennel and D. Schöder, "A predictive control strategy for converters," in Third IPAC Symposium, pp. 415–422, 1983.

[14] J. Rodríguez, J. Pontt, C. Silva *et al.* "Predictive direct torque control of an induction machine," EPE-PEMC (Power Electronics and Motion Control Conference), Riga, Latvia, September 2004.

[15] R. Kennel, J. Rodríguez, J. Espinoza, and M. Trincado, "High performance speed control methods for electrical machines: an assessment," in Industrial Technology (ICIT), pp. 1793–1799, 2010.

[16] K. P. Kovács, *Transient Phenomena in Electrical Machines*. Elsevier, 1984.

[17] J. Holtz, "The dynamic representation of AC drive systems by a complex signal flow graphs," in International Symposium on Industrial Electronics, ISIE'94, vol. 1, Santiago de Chile, pp. 1–6, 1994.

[18] J. Holtz, "The induction motor, a dynamic system," in IEEE IECON'94, vol. 1, Bologna, pp. 1–6, 1994.

[19] S. Muller, U. Ammann, and S. Rees, "New time-discrete modulation scheme for matrix converters," *IEEE Transactions on Industrial Electronics*, vol. 52, no. 6, pp. 1607–1615, December 2005.

[20] C. Klumpner, P. Nielsen, I. Boldea, and F. Blaabjerg, "A new matrix converter-motor (MCM) for industry applications," Conference Record of the IEEE Industry Applications Society Annual Meeting, IEEE-IAS, 2000.

[21] K.-B. Lee and F. Blaabjerg, "Improved sensorless vector control for induction motor drives fed by a matrix converter using nonlinear modeling and disturbance observer," *IEEE Transactions on Energy Conversion*, vol. 21, no. 1, pp. 52–59, March 2006.

[22] P. Wheeler, J. Rodríguez, J. Clare, L. Empringham, and A. Weinstein, "Matrix converters: a technology review," *IEEE Transactions on Industrial Electronics*, vol. 49, no. 2, pp. 276–288, April 2002.

9

Predictive Control of Permanent Magnet Synchronous Motors

9.1 Introduction

Permanent magnet synchronous motors (PMSMs) present several characteristics that make them very attractive for drive applications, such as high torque, high power density and efficiency, and excellent dynamic response. Because of these characteristics, PMSMs are suitable for a wide variety of applications including general purpose industrial drives, high-performance servo drives, and several specific applications where size and weight are restricted, as in automotive and aerospace applications.

PMSMs are composed, in general, of three-phase stator windings and an iron rotor with permanent magnets attached to it. The permanent magnets can be mounted on the rotor surface or inside the rotor core. In this way, the magnetic field is fixed to the rotor position. Due to its construction, the rotor's speed is rigidly related to the stator frequency and for variable speed operation a voltage source inverter is required.

Several control schemes have been proposed for the PMSM. Well-established methods are field-oriented control (FOC) [1] and direct torque control (DTC) [2, 3]. The quality of the FOC scheme depends on the performance of the current controllers, the most common being the use of PI controllers with PWM. Other control schemes like hysteresis and deadbeat-based controllers [4, 5] have also been proposed. The use of MPC for current control in a PMSM is presented in this chapter, based on similar control schemes that were reported in [6, 7].

A very different approach for the use of MPC in a PMSM drive considers direct control of the speed, as reported in [8] and explained further in this chapter.

9.2 Machine Equations

A PMSM with three-phase stator windings and a sinusoidal flux distribution is considered in this chapter. The machine is fed by a three-phase inverter as shown in Figure 9.1.

Predictive Control of Power Converters and Electrical Drives, First Edition. Jose Rodriguez and Patricio Cortes.

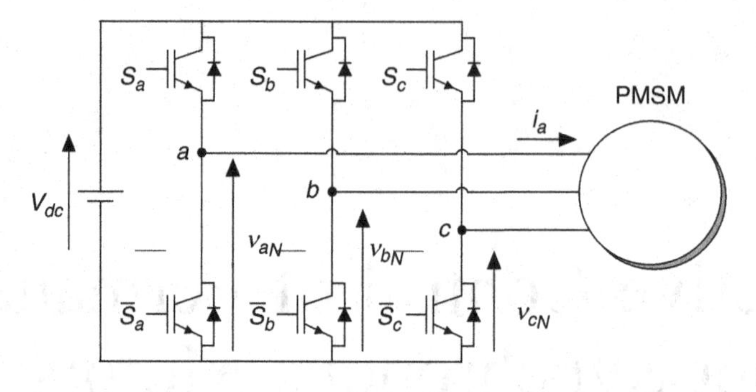

Figure 9.1 PMSM fed by a three-phase inverter

By defining the following space vector definitions for the stator voltage, stator current, and stator flux, respectively,

$$\mathbf{v}_s = \frac{2}{3}(v_{sa} + \mathbf{a}v_{sb} + \mathbf{a}^2 v_{sc}) \tag{9.1}$$

$$\mathbf{i}_s = \frac{2}{3}(i_{sa} + \mathbf{a}i_{sb} + \mathbf{a}^2 i_{sc}) \tag{9.2}$$

$$\boldsymbol{\psi}_s = \frac{2}{3}(\psi_{sa} + \mathbf{a}\psi_{sb} + \mathbf{a}^2 \psi_{sc}) \tag{9.3}$$

the stator dynamics can be described as

$$\mathbf{v}_s = R_s \mathbf{i}_s + \frac{d\boldsymbol{\psi}_s}{dt} \tag{9.4}$$

where R_s is the stator resistance.

The stator flux linkage $\boldsymbol{\psi}_s$ is generated by the rotor magnets and the self-linked flux produced by the stator currents. This relation is described by

$$\boldsymbol{\psi}_s = L_s \mathbf{i}_s + \psi_m e^{j\theta_r} \tag{9.5}$$

where L_s is the stator self-inductance, ψ_m is the flux magnitude due to the rotor magnets, and θ_r is the rotor position.

Inserting (9.5) into (9.4) we obtain

$$\mathbf{v}_s = R_s \mathbf{i}_s + L_s \frac{d\mathbf{i}_s}{dt} + j\psi_m \omega_r e^{j\theta_r} \tag{9.6}$$

where $\omega_r = d\theta_r/dt$ is the rotor speed.

Multiplying by $e^{-j\theta_r}$ and considering the stator voltage and current space vectors in rotor coordinates aligned with the rotor axis $\mathbf{v}_s^{(r)} = \mathbf{v}_s e^{-j\theta_r}$ and $\mathbf{i}_s^{(r)} = \mathbf{i}_s e^{-j\theta_r}$, (9.6) becomes

$$\mathbf{v}_s^{(r)} = R_s \mathbf{i}_s^{(r)} + L_s \frac{d\mathbf{i}_s^{(r)}}{dt} + jL_s \omega_r \mathbf{i}_s^{(r)} + j\psi_m \omega_r \tag{9.7}$$

where superscript (r) denotes rotor coordinates.

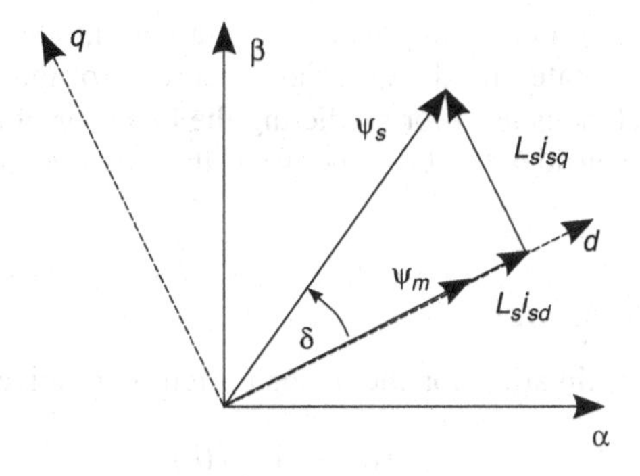

Figure 9.2 Vector diagram of the stator variables and the rotating reference frame

The stator equation (9.7) can be rewritten in dq coordinates

$$v_{sd} = R_s i_{sd} + L_s \frac{di_{sd}}{dt} - L_s \omega_r i_{sq} \tag{9.8}$$

$$v_{sq} = R_s i_{sq} + L_s \frac{di_{sq}}{dt} + L_s \omega_r i_{sd} + j\psi_m \omega_r \tag{9.9}$$

where $\mathbf{v}_s^{(r)} = v_{sd} + jv_{sq}$ and $\mathbf{i}_s^{(r)} = i_{sd} + ji_{sq}$.

The relation between the stator vectors and the rotating reference frame is shown in Figure 9.2.

The electric torque produced by the machine depends on the flux magnitude and the quadrature component of the stator current vector

$$T_e = \frac{3}{2} p \psi_m i_{sq} \tag{9.10}$$

The mechanical rotor dynamics are described by the equation

$$\frac{d\omega_r^m}{dt} = \frac{1}{J}(T_e - T_l) - \frac{B}{J}\omega_r^m \tag{9.11}$$

where ω_r^m is the rotor shaft's mechanical speed, J is the rotor inertia, B is the friction coefficient, and T_l is the load torque. The relation between the mechanical speed and electrical angular frequency is given by

$$\omega_r = p\omega_r^m \tag{9.12}$$

where p is the number of pole pairs of the machine.

9.3 Field Oriented Control Using Predictive Current Control

By using a rotating dq reference frame oriented to the rotor magnetic field axis, each stator current component has a physical meaning. The imaginary component i_{sq} is proportional to the electrical torque while the real component i_{sd} is proportional to the reactive power.

In this way, the machine control is implemented as a current control scheme, where the current references are generated by the external speed control loop.

The model of the machine is used for predicting the behavior of the stator currents, and the cost function must consider the error between the reference currents and predicted currents.

9.3.1 Discrete-Time Model

By using the Euler approximation for the stator current derivatives for a sampling time T_s, that is,

$$\frac{di}{dt} \approx \frac{i(k+1) - i(k)}{T_s}$$

the following expressions for the predicted stator currents in the dq reference frame are obtained from (9.8) and (9.9):

$$i_{sd}^p(k+1) = \left(1 - \frac{R_s T_s}{L_s}\right) i_{sd}(k) + T_s \omega_r i_{sq}(k) + \frac{T_s}{L_s} v_{sd} \tag{9.13}$$

$$i_{sq}^p(k+1) = \left(1 - \frac{R_s T_s}{L_s}\right) i_{sq}(k) - T_s \omega_r i_{sd}(k) - \psi_m \omega_r T_s + \frac{T_s}{L_s} v_{sq} \tag{9.14}$$

These equations allow predictions of the stator currents to be calculated for each one of the seven voltage vectors generated by the inverter. The voltage vectors generated by the inverter are fixed in the stationary reference frame, but they are rotating vectors in the dq reference frame, calculated as

$$\mathbf{v}_s^{(r)} = \mathbf{v}_s e^{-j\theta_r} \tag{9.15}$$

9.3.2 Control Scheme

The control scheme for FOC of the PMSM using predictive current control is shown in Figure 9.3. Here, a PI controller is used for speed control and generates the reference for the torque-producing current i_{sq}^*. A predictive current controller is used for tracking this current. In the predictive scheme, the discrete-time model of the machine is used for predicting the stator current components for the seven different voltage vectors generated by the inverter. The voltage vector that minimizes a cost function is selected and applied during a whole sampling interval.

The objectives of the predictive current control scheme are summarized as follows:

- Torque current reference tracking
- Torque by ampere optimization
- Current magnitude limitation.

These objectives can be expressed as the following cost function:

$$g = \left(i_{sd}^p(k+1)\right)^2 + (i_{sq}^* - i_{sq}^p(k+1))^2 + \hat{f}(i_{sd}^p(k+1), i_{sq}^p(k+1)) \tag{9.16}$$

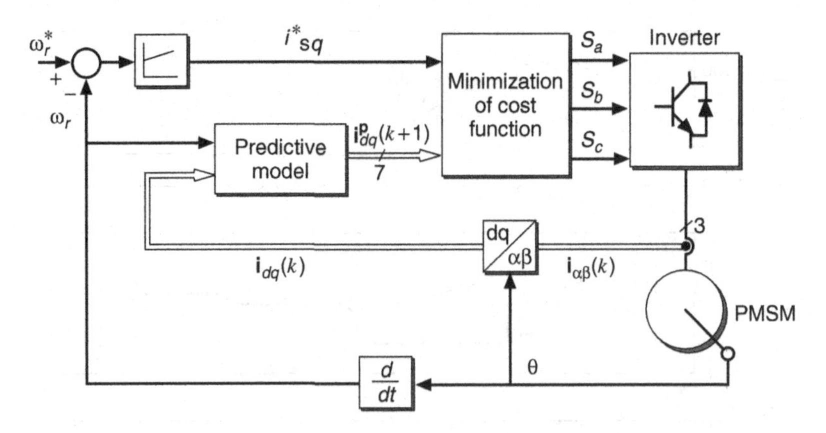

Figure 9.3 FOC of a PMSM using predictive current control (Fuentes *et al.*, 2009 © IEEE)

where the first term represents the minimization of the reactive power, allowing the torque by ampere optimization, the second term is defined for tracking the torque-producing current, and the last term is a nonlinear function for limiting the amplitude of the stator currents. This function is defined as

$$\hat{f}(i_{sd}^{p}(k+1), i_{sq}^{p}(k+1)) = \begin{cases} \infty \ \text{ if } |i_{sd}^{p}| > i_{max} \text{ or } |i_{sd}^{p}| > i_{max} \\ 0 \ \text{ if } |i_{sd}^{p}| \leq i_{max} \text{ and } |i_{sd}^{p}| \leq i_{max} \end{cases} \tag{9.17}$$

where i_{max} is the value of the maximum allowed stator current magnitude. In this way, if a given voltage vector generates predicted currents with a magnitude higher than i_{max} then the cost function will be $g = \infty$, and, in consequence, this voltage vector will not be selected. On the other hand, if the predicted stator currents are below the limits, the cost function will be composed of the first two terms only and the voltage vector that minimizes the current error will be selected.

Results using the predictive current control scheme are shown in Figure 9.4. In this figure a speed reference step change is performed at time $t = 0.02$ s, then a speed reversal at time $t = 0.1$ s, and finally a load step is applied at time 0.25 s. It can be seen from these results that all the objectives of the control are achieved during the tests. Fast tracking of the torque-producing current i_{sq} is achieved while the i_{sd} current is near zero. During transients the magnitude of the currents is limited and both components are decoupled.

The behavior of the predictive current control for a speed reversal is shown in Figure 9.5. During this transient the maximum allowed i_{sq} current is applied, which is effectively limited by the last term of the cost function (9.16). The i_{sd} current component is near zero even during the transients.

A step change in the load was applied by a loading machine coupled to the shaft of the PMSM. The behavior of the machine variables for this test is shown in Figure 9.6. The speed PI controller responds to this disturbance by changing the torque current reference i_{sq}, which is followed by the predictive current control.

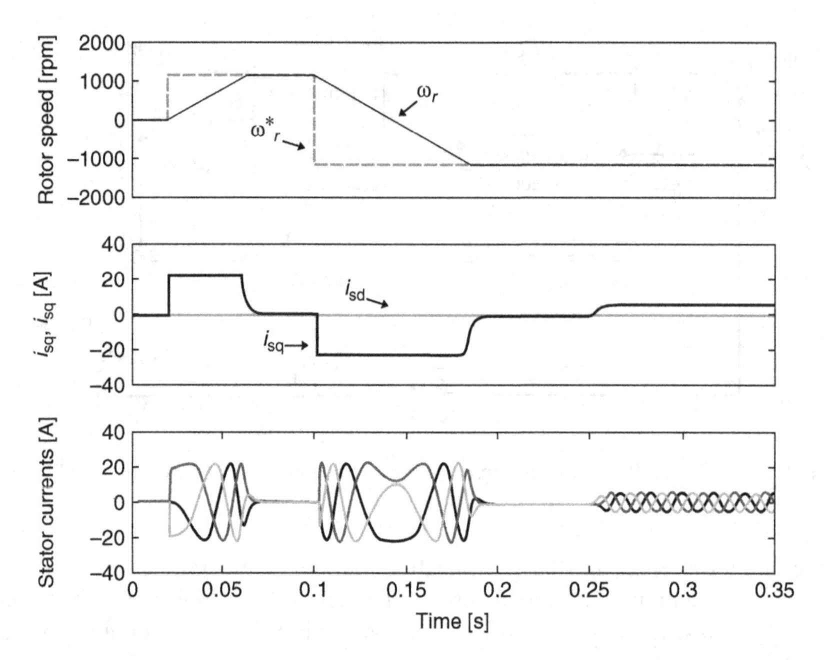

Figure 9.4 Behavior of the predictive current control of a PMSM. Rotor speed and stator currents (Fuentes *et al.*, 2009 © IEEE)

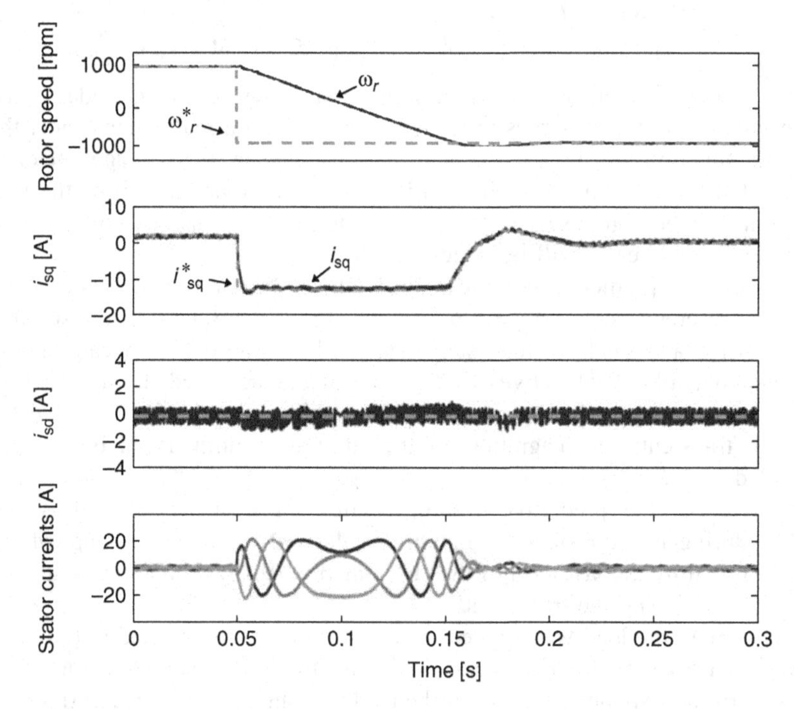

Figure 9.5 Behavior of the rotor speed and stator currents of the PMSM for a speed reversal (Fuentes *et al.*, 2009 © IEEE)

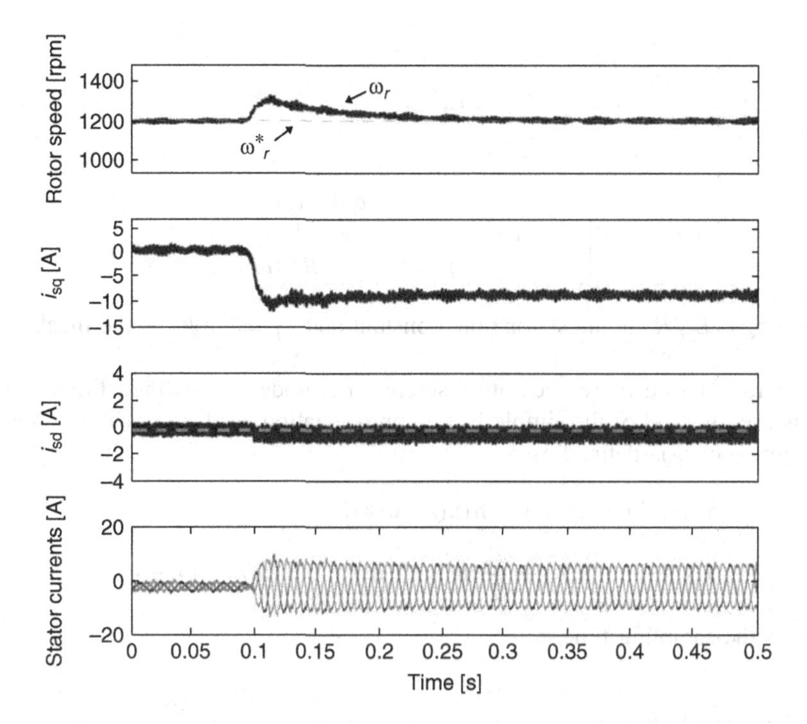

Figure 9.6 Behavior of the rotor speed and stator currents for a load torque step change (Fuentes *et al.*, 2009 © IEEE)

9.4 Predictive Speed Control

One of the main advantages of MPC is the possibility to control several different variables using a single cost function. This makes possible the implementation of a predictive speed control while maintaining the stator currents under given conditions. It is desirable to consider a limitation on the magnitude of the currents and to optimize the torque by ampere ratio.

This application of MPC introduces some difficulties, such as the nonlinear model of the system, the large differences between the speed dynamics (mechanical dynamics) and the dynamics of the stator currents (electrical dynamics), and the quantization noise in the speed measurement [8].

9.4.1 Discrete-Time Model

In order to perform a speed control using MPC, the mechanical equations must be included in the discrete-time equations used for calculation of the predictions. The machine model can be summarized in the form of the following space state equation:

$$\frac{d\mathbf{x}}{dt} = h(\mathbf{x}, \mathbf{u}) \tag{9.18}$$

where

$$\mathbf{x} = [i_{sd} \quad i_{sq} \quad \omega_r]^T \tag{9.19}$$

$$\mathbf{u} = [v_{sd} \quad v_{sq}]^T \tag{9.20}$$

$$h(\mathbf{x}, \mathbf{u}) = \begin{bmatrix} -1/\tau_s i_{sd} + \omega_r i_{sq} + 1/L_s v_{sd} \\ -1/\tau_s i_{sq} - \omega_r i_{sd} - \psi_m/L_s \omega_r + 1/L_s v_{sq} \\ pk_T/J i_{sq} - B/J\omega_r \end{bmatrix} \tag{9.21}$$

and where $\tau_s = L_s/R_s$ is the stator time constant and $k_T = \frac{3}{2}p\psi_m$ is the machine torque constant.

In order to obtain a more accurate discrete-time model, a modified Euler integration method is used instead of the simple Euler approximation used in previous chapters. The discrete-time model is defined as

$$\hat{\mathbf{x}}(k+1) = \mathbf{x}(k) + T_s h(\mathbf{x}(k), \mathbf{u}(k)) \tag{9.22}$$

$$\mathbf{x}(k+1) = \mathbf{x}(k) + \frac{T_s}{2}(h(\mathbf{x}(k), \mathbf{u}(k)) + h(\hat{\mathbf{x}}(k+1), \mathbf{u}(k))) \tag{9.23}$$

where T_s is the sampling time.

9.4.2 Control Scheme

A block diagram of the predictive speed control is shown in Figure 9.7. The discrete-time model of the machine is used for calculating predictions of the rotor speed and stator currents for the seven voltage vectors generated by the inverter. These predictions are evaluated by a cost function that defines the desired behavior of the system. The voltage vector that minimizes this function is selected and applied to the machine terminals for a whole sampling interval.

Due to the noisy nature of the speed measurement and the high sampling frequencies that are usually needed in this kind of predictive control scheme, the use of an extended Kalman filter (EKF) was proposed in [8] for estimating the rotor speed.

The objectives of the predictive speed control scheme are summarized as follows:

- Speed reference tracking
- Smooth behavior of the electrical torque
- Torque by ampere optimization
- Current magnitude limitation.

These objectives can be expressed as the following cost function:

$$g = \lambda_\omega(\omega_r^* - \omega_r^p(k+1))^2 + \lambda_i\left(i_{sd}^p(k+1)\right)^2 + \lambda_{if}(i_{sqf}^p(k+1))^2 + \hat{f}(i_{sd}^p(k+1), i_{sq}^p(k+1)) \tag{9.24}$$

where the first term evaluates the predicted speed error and ω_r^* is the speed reference. The second term minimizes the i_{sd} current for optimized torque by ampere ratio. The third term

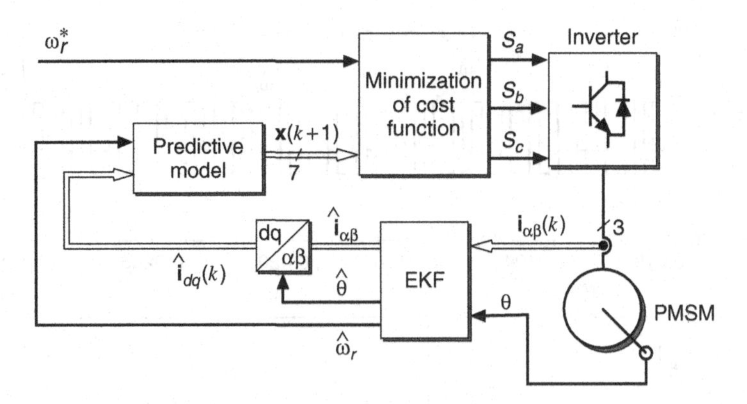

Figure 9.7 Predictive speed control of a PMSM

evaluates the filtered value of the i_{sq} current. A high-pass filter is used for penalization of the high-frequency content of the torque-producing current in order to obtain a smooth behavior for the electrical torque. The last term is a nonlinear function, as defined in (9.17), that allows limitation of the stator currents by penalizing the voltage vectors that make the predicted current magnitude higher than the defined limits. Considering that the different terms of this cost function have different units and magnitudes, weighting factors λ_{ω}, λ_i, and λ_{if} must be introduced. These weighting factors allow compensation of these differences and are design parameters that can be used for tuning the controller.

9.4.3 Rotor Speed Estimation

An incremental encoder is considered for rotor position measurement. These devices give a quantized measurement of the rotor position, introducing a high-frequency noise into the angle measurement. As the rotor speed is the derivative of the rotor angle, the quantization noise is amplified in the speed measurement. The rotor speed can be calculated using the Euler approximation of the derivative, for a sampling time T_s:

$$\tilde{\omega}_r(k) = \frac{\theta_r(k) - \theta_r(k-1)}{T_s} \tag{9.25}$$

It can be seen from this equation that the energy of the noise increases if the resolution of the encoder is low or the sampling frequency is high.

This high-frequency noise is not a problem in classical control schemes due to the low-pass nature of the PI controllers that are commonly used for the speed control loop. However, in predictive speed control, this high level of high-frequency noise impedes the correct operation of the control strategy. In order to overcome this problem, the use of an extended Kalman filter (EKF) was proposed in [8] for estimating the rotor speed.

The EKF is implemented using the standard method presented in [9].

The control strategy was implemented using a sampling time of $T_s = 30\,\mu\text{s}$. The values of the weighting factors used for these results are: $\lambda_{\omega} = 1000$, $\lambda_i = 1$, and $\lambda_{if} = 1.4$. The high-pass filter for the i_{sq} current is a second-order Butterworth filter with a cutoff frequency $f_c = 200\,\text{Hz}$. The current limitation considered in (9.17) is $i_{max} = 22.6\,\text{A}$.

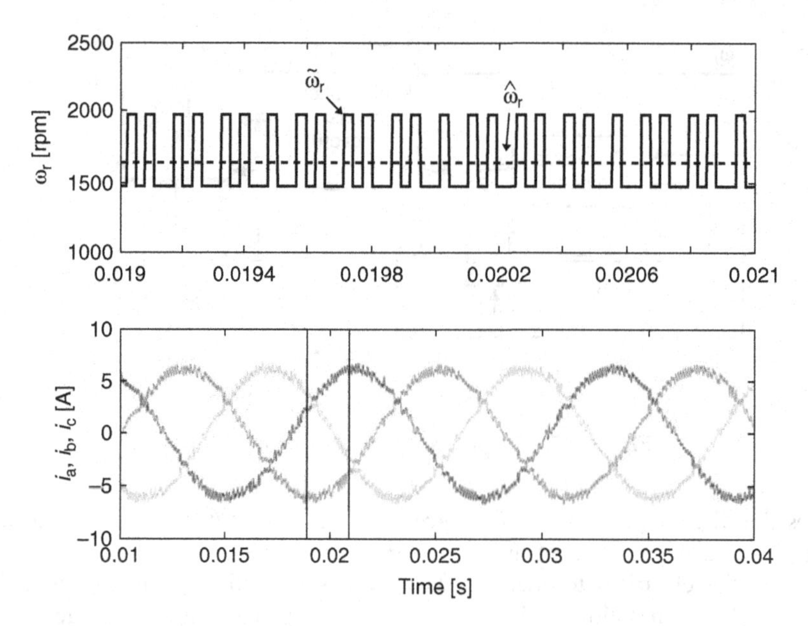

Figure 9.8 Steady state operation of the rotor speed and stator currents of the PMSM (Fuentes *et al.*, 2009 © IEEE)

Results at steady state operation are shown in Figure 9.8. The effect of using the EKF is observed in this figure, where estimated speed $\hat{\omega}_r$ does not present the noise due to quantization, which is very evident in the waveform of the speed obtained using the Euler approximation for the derivative of the rotor position $\tilde{\omega}_r$. An incremental encoder with a resolution of 4096 pulses per revolution sampled at $T_s = 30\,\mu\text{s}$ is considered for these results. The stator currents are sinusoidal and present low distortion as a consequence of the minimization of i_{sd}.

The behavior of the predictive speed control for a speed reversal is shown in Figure 9.9. It can be observed that the speed response presents a nearly ideal behavior with good reference tracking and almost no overshoot. The torque-producing current i_{sq} is adjusted with fast dynamics in order to obtain the desired speed behavior, while the i_{sd} current is almost zero, even during transients. This result shows that both current components are highly decoupled. The stator currents are sinusoidal during all the tests due to the minimization of i_{sd}.

9.5 Summary

This chapter presents the application of MPC for the control of permanent magnet synchronous motors. Two control schemes are presented, a field-oriented control scheme with a predictive current controller and a predictive speed control scheme.

The predictive current control is similar to the one presented in Chapter 4, but in this case it is implemented in the rotating reference frame. In this way, one of the current components is proportional to the electrical torque and the other is proportional to the reactive power. The current references are generated by an external speed control loop.

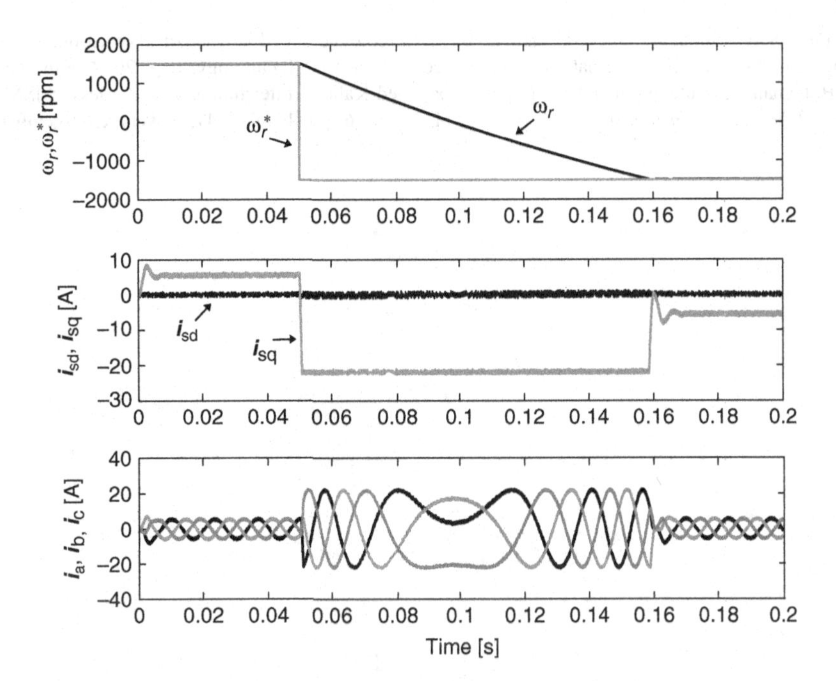

Figure 9.9 Behavior of the rotor speed and stator currents of the PMSM for a speed reversal (Fuentes *et al.*, 2009 © IEEE)

The predictive speed control does not use any external speed control loop and the speed is directly controlled by the predictive controller. This application is a very good example of how different variables and different control objectives and restrictions can be included in the MPC scheme, resulting in a high dynamic performance of the system.

References

[1] T. M. Jahns, G. B. Kliman, and T. W. Neumann, "Interior permanent-magnet synchronous motors for adjustable-speed drives," *IEEE Transactions on Industry Applications*, vol. IA-22, no. 4, pp. 738–747, July 1986.

[2] C. French and P. Acarnley, "Direct torque control of permanent magnet drives," *IEEE Transactions on Industry Applications*, vol. 32, no. 5, pp. 1080–1088, September/October 1996.

[3] L. Tang, L. Zhong, M. Rahman, and Y. Hu, "A novel direct torque control for interior permanent-magnet synchronous machine drive with low ripple in torque and flux: a speed-sensorless approach," *IEEE Transactions on Industry Applications*, vol. 39, no. 6, pp. 1748–1756, March 2003.

[4] H. Le-Huy, K. Slimani, and P. Viarouge, "Analisis and implementation of a real-time predictive current controller for permanent-magnet synchronous servo drives," *IEEE Transactions on Industrial Electronics*, vol. 41, no. 1, pp. 110–117, February 1994.

[5] H.-T. Moon, H.-S. Kim, and M.-J. Youn, "A discrete-time predictive current control for PMSM," *IEEE Transactions on Power Electronics*, vol. 18, no. 1, pp. 464–472, January 2003.

[6] E. Fuentes, J. Rodríguez, C. Silva, S. Diaz, and D. Quevedo, "Speed control of a permanent magnet synchronous motor using predictive current control," in IEEE 6th International Power Electronics and Motion Control Conference, IPEMC '09, pp. 390–395, May 2009.

[7] F. Morel, X. Lin-Shi, J.-M. Retif, B. Allard, and C. Buttay, "A comparative study of predictive current control schemes for a permanent-magnet synchronous machine drive," *IEEE Transactions on Industrial Electronics*, vol. 56, no. 7, pp. 2715–2728, July 2009.

[8] E. Fuentes, C. Silva, D. Quevedo, and E. Silva, "Predictive speed control of a synchronous permanent magnet motor," in IEEE International Conference on Industrial Technology, ICIT 2009, 2009, pp. 1–6.
[9] S. Bolognani, L. Tubiana, and M. Zigliotto, "Extended Kalman filter tuning in sensorless PMSM drives," *IEEE Transactions on Industry Applications*, vol. 39, no. 6, pp. 1741–1747, November/December 2003.

Part Four

Design and Implementation Issues of Model Predictive Control

10

Cost Function Selection

10.1 Introduction

This chapter introduces several types of terms that can be included in a cost function and shows how these terms are related to different control requirements for the system.

10.2 Reference Following

The most common terms in a cost function are the ones that represent a variable following a reference. Some examples are current control, torque control, power control, etc. These terms can be expressed in a general way as the error between the predicted variable and its reference:

$$g = ||x^* - x^p|| \tag{10.1}$$

where x^* is the reference value and x^p is the predicted value of the controlled variable, for a given switching state of the power converter. The norm $||\cdot||$ is a measure of distance between reference and predicted values and usually it can be implemented as an absolute value, square value, or integral value of the error for one sampling period:

$$g = |x^* - x^p| \tag{10.2}$$

$$g = (x^* - x^p)^2 \tag{10.3}$$

$$g = \left| \int_k^{k+1} (x^*(t) - x^p(t))\, dt \right| \tag{10.4}$$

Absolute error and squared error give similar results when the cost function considers only one error term. However, if the cost function has two or more different terms, results can be different. As will be shown in the next section, squared error presents a better reference following when additional terms are included in the cost function. Cost function (10.4) considers the trajectory of the variable between time t_k and t_{k+1}, not just the final value at instant t_{k+1}, leading to the mean value of the error to be minimized. This then leads to more accurate reference tracking.

Predictive Control of Power Converters and Electrical Drives, First Edition. Jose Rodriguez and Patricio Cortes.

10.2.1 Some Examples

This simple type of cost function can be used to control several systems like those presented in this book. Current control can be implemented for three-phase systems using the following cost function expressed in orthogonal coordinates:

$$g = \left|i_\alpha^* - i_\alpha^p\right| + \left|i_\beta^* - i_\beta^p\right| \tag{10.5}$$

which was applied to a three-phase inverter in Chapter 4 and [1], to an active front-end rectifier in Chapter 6 and [2], and to a matrix converter in Chapter 7 and [3].

Direct power control can be achieved using this type of cost function

$$g = \left|P^* - P^p\right| + \left|Q^* - Q^p\right| \tag{10.6}$$

as presented in Chapter 6.

Another variation of these types of cost functions includes reference following of two variables with different magnitudes. Such is the case for the predictive torque and flux control presented in Chapter 8, which considers the following cost function:

$$g = \left|T_e^* - T_e^p\right| + \lambda_\psi \left||\boldsymbol{\psi}|^* - |\boldsymbol{\psi}|^p\right| \tag{10.7}$$

where λ_ψ is a weighting factor that handles the difference in magnitude of the two reference following terms. This factor can be also adjusted in order to modify the importance of each term, as will be explained in the next chapter.

10.3 Actuation Constraints

In a control system it is important to reach a compromise between reference following and control effort. In power converters and drives, the control effort is related to the voltage or current variations, the switching frequency, or the switching losses. Using predictive control, it is possible to consider any measure of control effort in the cost function, in order to reduce it.

In a three-phase inverter, the control effort can be represented by the change in the voltage vector applied to the load. This can be implemented as an additional term in the cost function measuring the magnitude of the difference between the previously applied voltage vector $\mathbf{v}(k-1)$ and the voltage vector to be applied $\mathbf{v}(k)$, resulting in

$$g = \left\|x^* - x^p\right\| + \lambda \left\|\mathbf{v}(k-1) - \mathbf{v}(k)\right\| \tag{10.8}$$

where x is the controlled variable and λ is a weighting factor that allows the level of compromise to be adjusted between reference following and control effort.

As an example, the predictive current control presented in Chapter 4 is considered, together with the constraint in the variation of the voltage vectors presented in (10.8). The resulting cost function is expressed as

$$g = \left|i_\alpha^* - i_\alpha^p\right| + \left|i_\beta^* - i_\beta^p\right| + \lambda \left\|\mathbf{v}(k-1) - \mathbf{v}(k)\right\| \tag{10.9}$$

By using this cost function, the control effort can be reduced by increasing the value of the weighting factor λ. Results for the current control of a three-phase inverter using cost function (10.9) are shown in Figure 10.1. It can be seen that by reducing the voltage

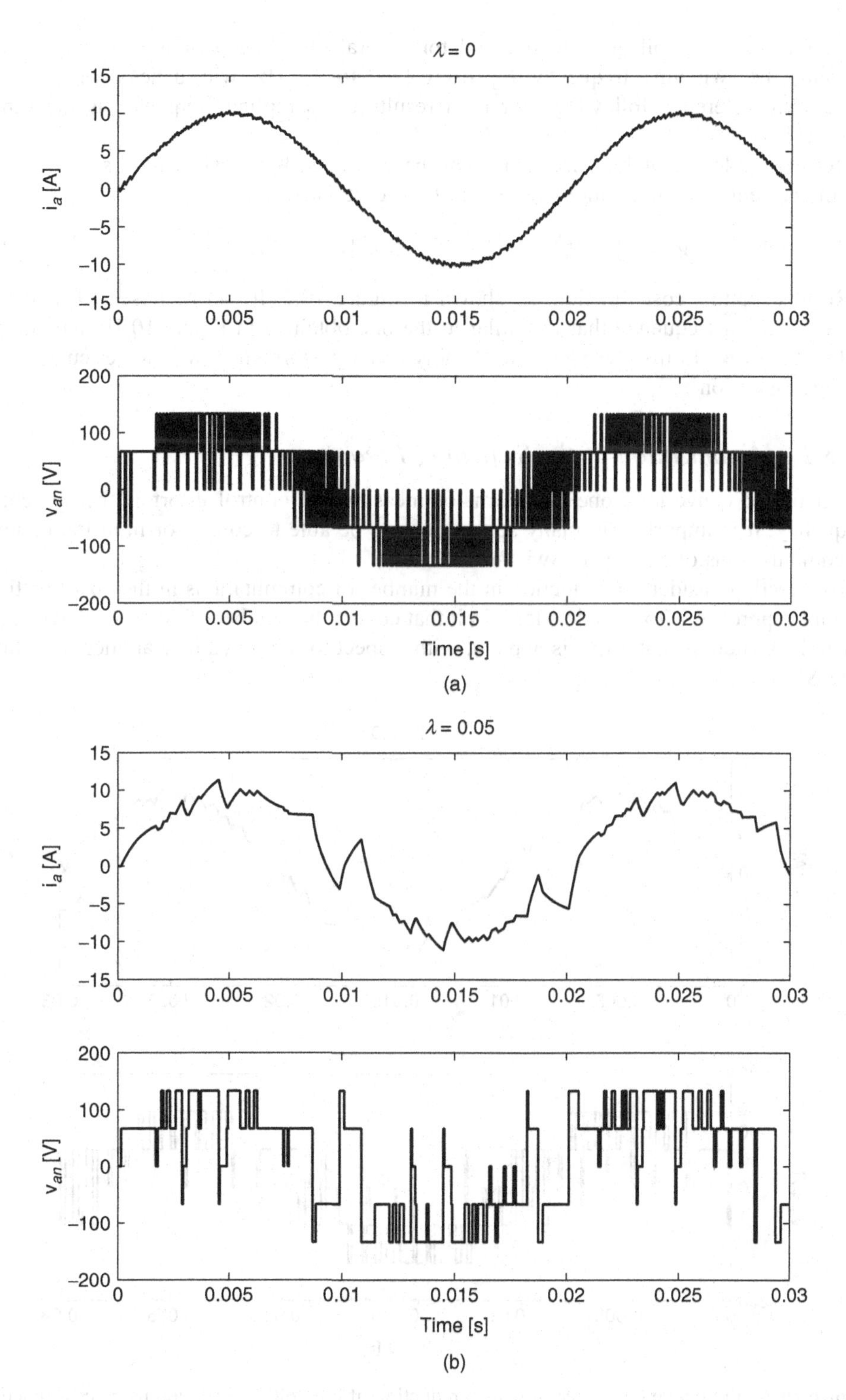

Figure 10.1 Predictive current control with reduction of the voltage variation using cost function (10.9). (a) Switching frequency $f_{sw} = 2767\,\text{Hz}$. (b) Switching frequency $f_{sw} = 723\,\text{Hz}$

variation, the load voltage is maintained for several sampling periods at a fixed value, reducing the switching frequency f_{sw} from 2767 to 723 Hz, with a negative effect on the current reference following. For both results the sampling frequency is the same, $f_s = 20\,\text{kHz}$.

Results for low switching frequency can be improved by considering a squared error for the current reference following terms of the cost function

$$g = \left(i_\alpha^* - i_\alpha^p\right)^2 + \left(i_\beta^* - i_\beta^p\right)^2 + \lambda \left\| \mathbf{v}(k-1) - \mathbf{v}(k) \right\| \tag{10.10}$$

Results for this cost function are shown in Figure 10.2. It can be observed here that for a switching frequency that is similar to the one obtained in Figure 10.1b, the current and voltage waveforms present a considerably better performance, with lower current and voltage distortion.

10.3.1 Minimization of the Switching Frequency

As in power converters, one of the major measures of control effort is the switching frequency. It is important in many applications to be able to control or limit the number of commutations of the power switches.

To directly consider the reduction in the number of commutations in the cost function, a simple approach is to include a term in it that covers the number of switches that change when the switching state $\mathbf{S}(k)$ is applied, with respect to the previously applied switching state $\mathbf{S}(k-1)$.

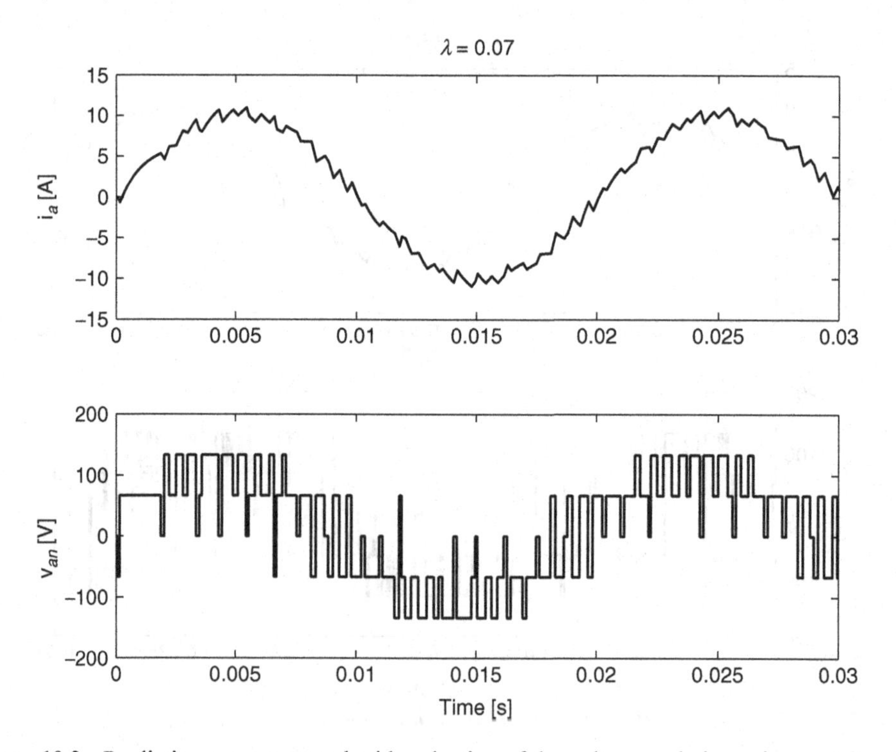

Figure 10.2 Predictive current control with reduction of the voltage variation using cost function (10.10). Switching frequency $f_{sw} = 725\,\text{Hz}$

The resulting cost function is expressed as

$$g = \left(i_\alpha^* - i_\alpha^p\right)^2 + \left(i_\beta^* - i_\beta^p\right)^2 + \lambda_n \cdot n \tag{10.11}$$

where n is the number of switches that change when the switching state $\mathbf{S}(k)$ is applied.

If the switching state vector $\mathbf{S}$ is defined as

$$\mathbf{S} = \left(S_1, S_2, \ldots, S_N\right) \tag{10.12}$$

where each element S_x represents the state of a switch and has only two states, one or zero, then the number of switches that change from $\mathbf{S}(k-1)$ to $\mathbf{S}(k)$ is

$$n = \sum_{x=1}^{N} \left|S_x(k) - S_x(k-1)\right| \tag{10.13}$$

Considering the three-phase inverter as an example, the switching state vector $\mathbf{S} = (S_a, S_b, S_c)$ defines the switching state of each inverter leg. Then the number of switches changing from time $k-1$ to time k is

$$n = \left|S_a(k) - S_a(k-1)\right| + \left|S_b(k) - S_b(k-1)\right| + \left|S_c(k) - S_c(k-1)\right| \tag{10.14}$$

The behavior of the switching frequency for different values of the weighting factor λ_n is shown in Figure 10.3. It can be seen that the switching frequency can be reduced to

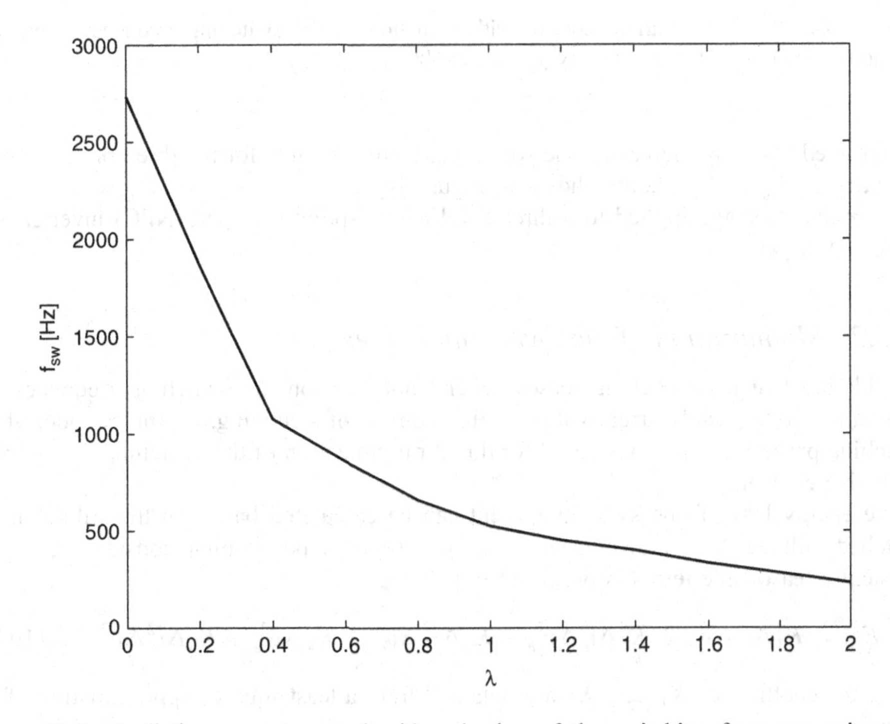

Figure 10.3 Predictive current control with reduction of the switching frequency using cost function (10.11)

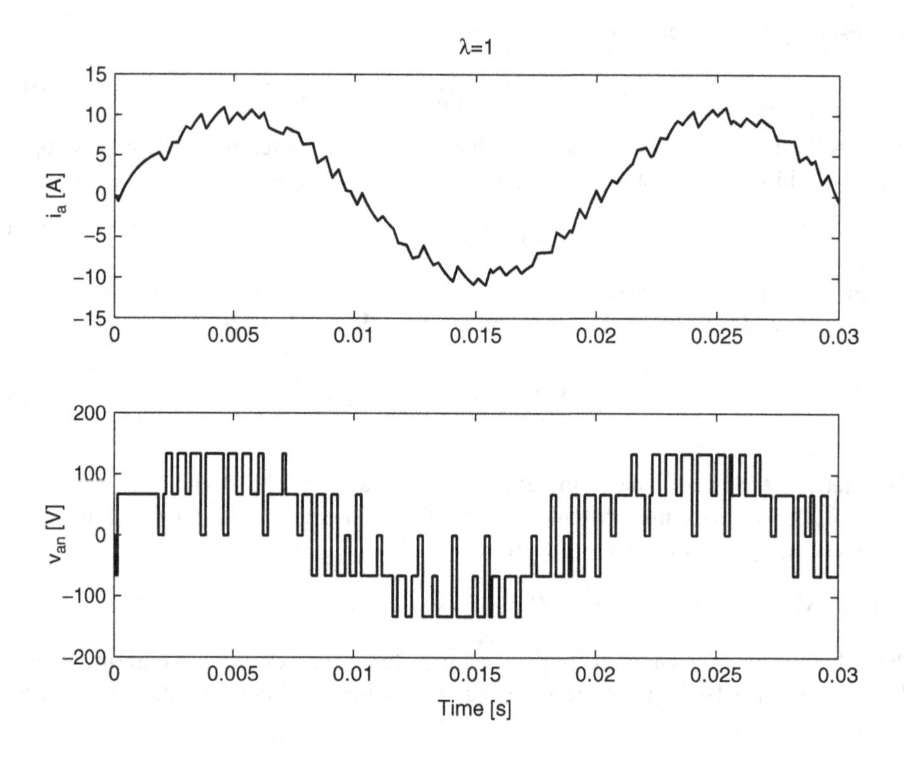

Figure 10.4 Predictive current control with reduction of the switching frequency using cost function (10.11). Switching frequency $f_{sw} = 525\,\text{Hz}$

the required value by increasing the weighting factor. Results for the three-phase inverter operating at $f_{sw} = 525\,\text{Hz}$ are shown in Figure 10.4.

A similar strategy, applied to a three-level neutral-point clamped (NPC) inverter, was presented in [4].

10.3.2 Minimization of the Switching Losses

Considering that the switching losses depend not only on the switching frequency but also on the voltage and current values at the moment of switching, a simple model of the switching process can be considered for direct minimization of the switching losses using predictive control.

The energy loss of one switching event can be calculated based on the values of the switched voltage Δv_{ce} and current Δi_c. The use of a polynomial considering all the physically reasonable terms is proposed in [5]:

$$E_{sw} = K_1 \Delta i_c \Delta v_{ce} + K_2 \Delta i_c \Delta v_{ce}^2 + K_3 \Delta i_c^2 \Delta v_{ce} + K_4 \Delta v_{ce}^2 + K_5 \Delta i_c^2 \Delta v_{ce}^2 \quad (10.15)$$

where the coefficients $K_1, \ldots, K_5$ are obtained from a least squares approximation of the measured data.

In order to obtain a simplified expression for estimating the losses, it is possible to neglect several of the terms in (10.15) by considering the experimental data, resulting in

$$E_{sw} = K_1 \Delta i_c \Delta v_{ce} \tag{10.16}$$

This expression is equivalent to the resulting equation obtained from considering the simplified current and voltage waveforms during the commutation process shown in Figure 10.5. From this figure it is possible to calculate the dissipated energy during the commutation process by integrating the instantaneous power:

$$E_{sw} = \int_{T_c} p(t)\, dt \tag{10.17}$$

or

$$E_{sw} = \int_{T_c} i_c(t) v_{ce}(t)\, dt = \frac{T_c}{6} \Delta i_c \Delta v_{ce} \tag{10.18}$$

where T_c is the duration of the commutation.

This simple model for estimating the switching losses can be included in the cost function as a term that also includes the predicted losses of all the switches of the power converter:

$$g = \left|\left|x^* - x^p\right|\right| + \lambda \sum_{j=1}^{N} \Delta i_{cp}(j) \Delta v_{cep}(j) \tag{10.19}$$

where N is the number of switching devices in the converter.

This control strategy for reducting the switching losses was proposed for a matrix converter in [6]. This work considers a cost function for output current control, input reactive power minimization, and reduction of switching losses:

$$g = \left|i^*_{o\alpha} - i^p_{o\alpha}\right| + \left|i^*_{o\beta} - i^p_{o\beta}\right| + A\left|Q^p\right| + B \sum_{j=1}^{18} \Delta i_{cp}(j) \Delta v_{cep}(j) \tag{10.20}$$

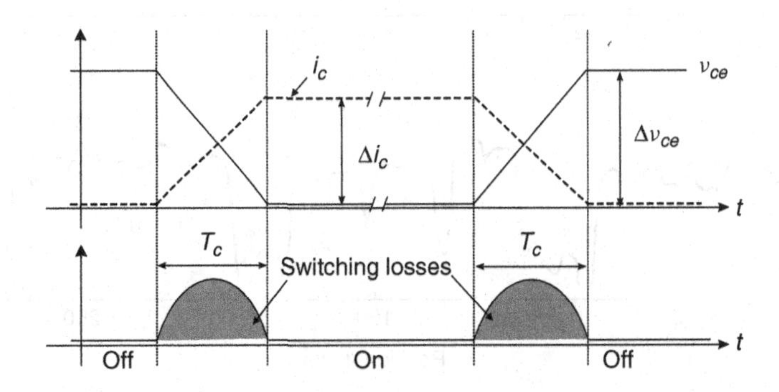

Figure 10.5 Simple model for estimation of the switching losses

where $\mathbf{i}_o = i_{o\alpha} + i_{o\beta}$ is the output current vector, Q is the input reactive power, and A and B are weighting factors. A detailed explanation of the predictive control of the matrix converter is presented in Chapter 7.

It was shown in [6] that by adjusting the value of B the efficiency of the matrix converter can be increased, maintaining the good performance, in terms of THD, of the input and output currents. By using a higher value of B the switching losses can be further reduced, but the THD of the currents will be increased. The temperature of the matrix converter switches, obtained using a thermal camera, is shown in Figure 10.6 for different values of B. It can be seen in the figure that as the value of B increases, the temperature of the switches decreases, as a result of lower switching losses.

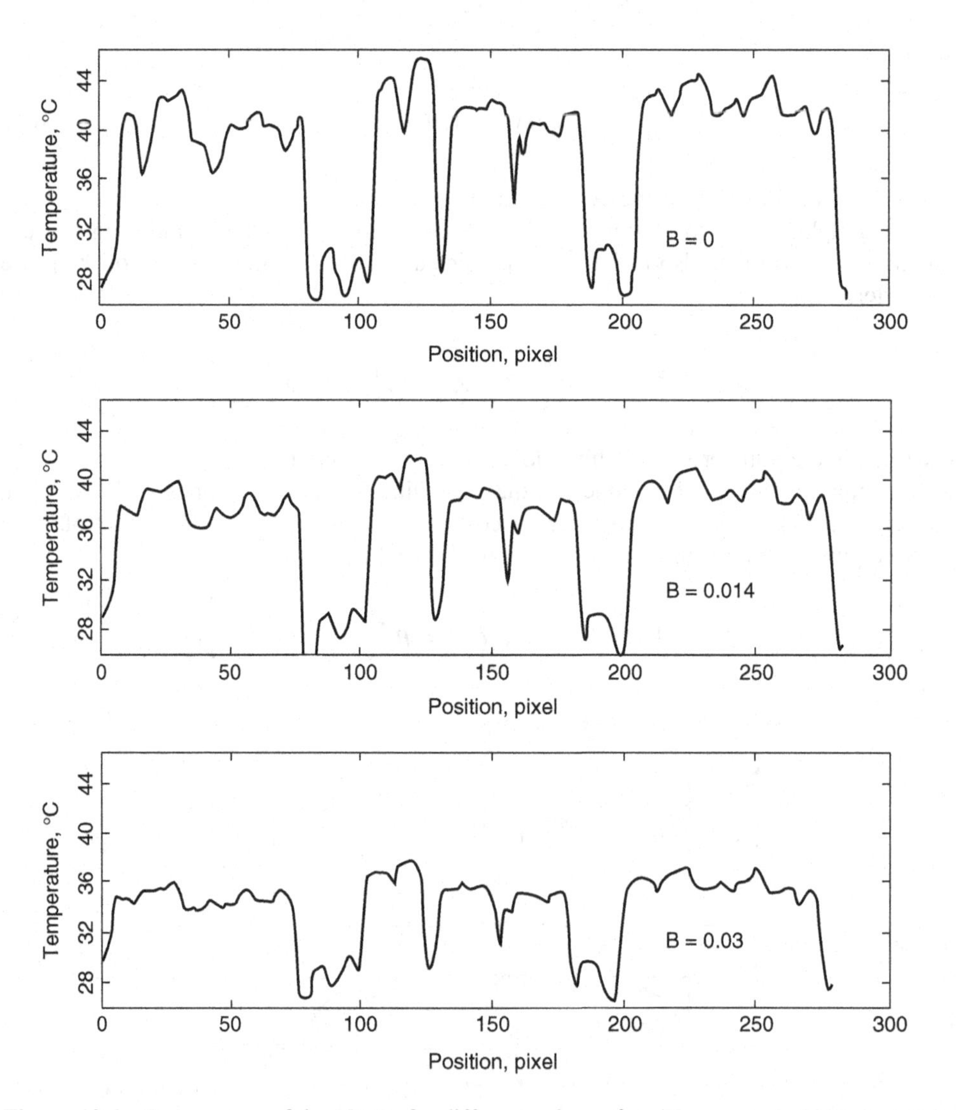

Figure 10.6 Temperature of the IGBTs for different values of B (Vargas *et al.*, 2008 © IEEE)

10.4 Hard Constraints

One of the advantages of predictive control is the possibility of achieving direct control of the output variables without the need for inner control loops. However, it can be found in several cases that as the internal variables are not controlled, they can reach values that are outside their allowed range. In a traditional cascaded control scheme, this kind of limitation of the internal variables is considered by including saturation levels for the references of these variables. In a predictive control scheme, these limitations can be included as an additional term in the cost function.

As an example, the predictive torque control explained in Chapter 8 will be considered. In this control scheme, the electrical torque T_e and the stator flux magnitude $|\psi_s|$ are directly controlled using the following cost function [7]:

$$g = \frac{(T_e^* - T_e)^2}{T_n^2} + A\frac{(|\psi_s|^* - |\psi_s|)^2}{\psi_{sn}^2} \tag{10.21}$$

Here, the stator currents are not directly controlled and at steady state they are sinusoidal and their magnitude is within the allowed limits. However, during some transients these currents can be very high, damaging the inverter or the machine. The startup of the induction machine using this control scheme is shown in Figure 10.7. It can be seen in

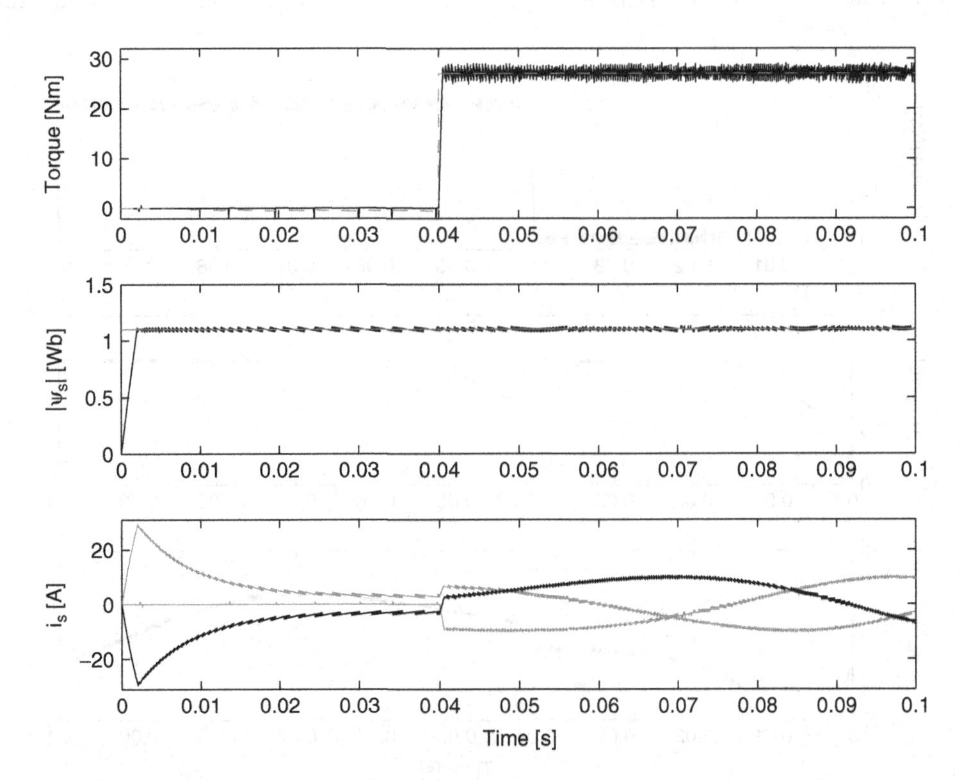

Figure 10.7 Predictive torque and flux control. Torque, stator flux magnitude, and stator currents at startup (Miranda *et al.*, 2009 © IEEE)

this figure that during the initial transient the stator currents can be considerably higher than the currents at full torque. It is desirable to consider the limitation of these currents in the predictive control scheme.

Taking into account the idea of predictive control, the optimization algorithm must discard any switching state that makes the predicted currents exceed the predefined limit, and from those that do not violate the limits select the one that minimizes the torque and flux error. This procedure can be implemented as an additional nonlinear term in the cost function that generates a very high value when the currents exceed the allowed limits, and is equal to zero when the currents are within the limits. The resulting cost function is

$$g = \frac{(T_e^* - T_e)^2}{T_n^2} + A\frac{\left(|\boldsymbol{\psi}_s|^* - |\boldsymbol{\psi}_s|\right)^2}{\psi_{sn}^2} + f_{lim}(\mathbf{i}_s^p) \tag{10.22}$$

where $\mathbf{i}_s^p$ is the predicted stator current vector, $f_{lim}(\mathbf{i}_s^p)$ is a nonlinear function defined as

$$f_{lim}(\mathbf{i}_s^p) = \begin{cases} \infty & \text{if} \left|\mathbf{i}_s^p\right| > i_{max} \\ 0 & \text{if} \left|\mathbf{i}_s^p\right| \leq i_{max} \end{cases} \tag{10.23}$$

and i_{max} is the value of the maximum allowed stator current magnitude.

The effect of this additional term in the cost function can be observed in Figure 10.8 for the same startup conditions as in Figure 10.7. It can also be seen that the stator current

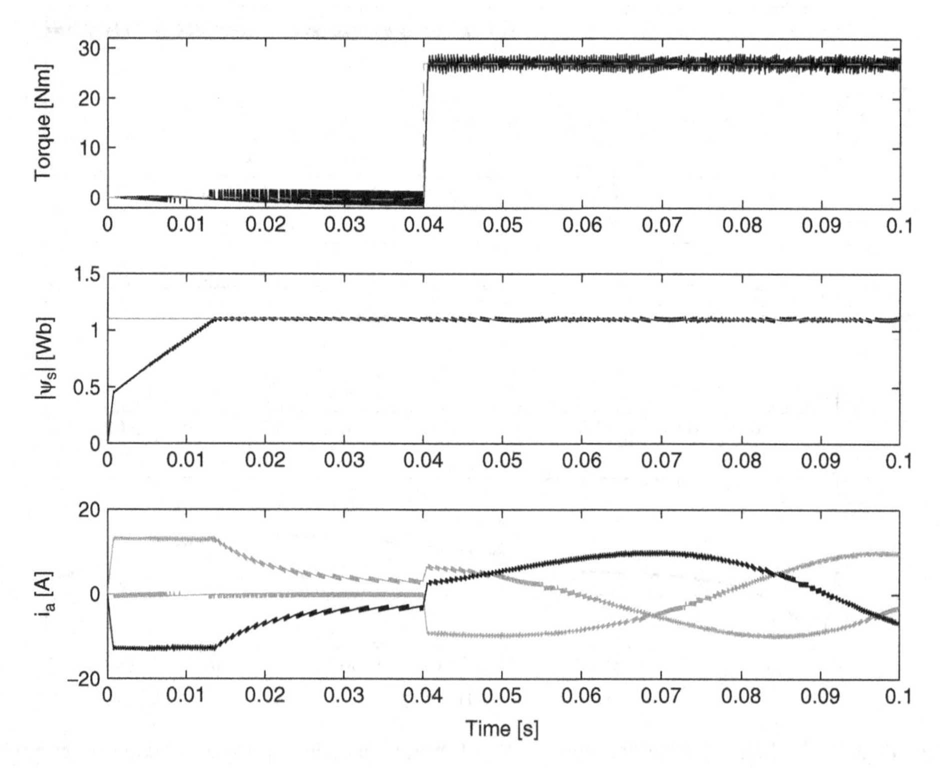

Figure 10.8 Predictive torque and flux control. Torque, stator flux magnitude, and stator currents at startup with stator current limitation (Miranda *et al.*, 2009 © IEEE)

magnitude is saturated at the defined limit $i_{max} = 15$ A. The operation of the predictive control when the stator currents are below the limits is identical for both cases.

Another example of current limitation can be found in Chapter 9 for a permanent magnet synchronous motor.

The same idea of using a nonlinear function like the one presented in this section can be used to limit any variable in any predictive control scheme.

10.5 Spectral Content

In addition to the control of the instantaneous values of the variables, it is possible to include requirements in the cost function for the spectral content of the variables.

The general implementation of predictive control presented in this book does not impose any pattern on the switching signals. The maximum switching frequency is limited by the sampling frequency but the optimal switching state could be maintained by several sampling periods. This results in variable switching frequency and a spread spectrum of the controlled variables.

Taking the predictive current control scheme presented in Chapter 4 as an example, it can be observed that the load currents display a spread spectrum like the ones shown in Figure 10.9. Here, two different sampling frequencies have been considered, presenting different ranges and magnitudes of the spectral content in each case.

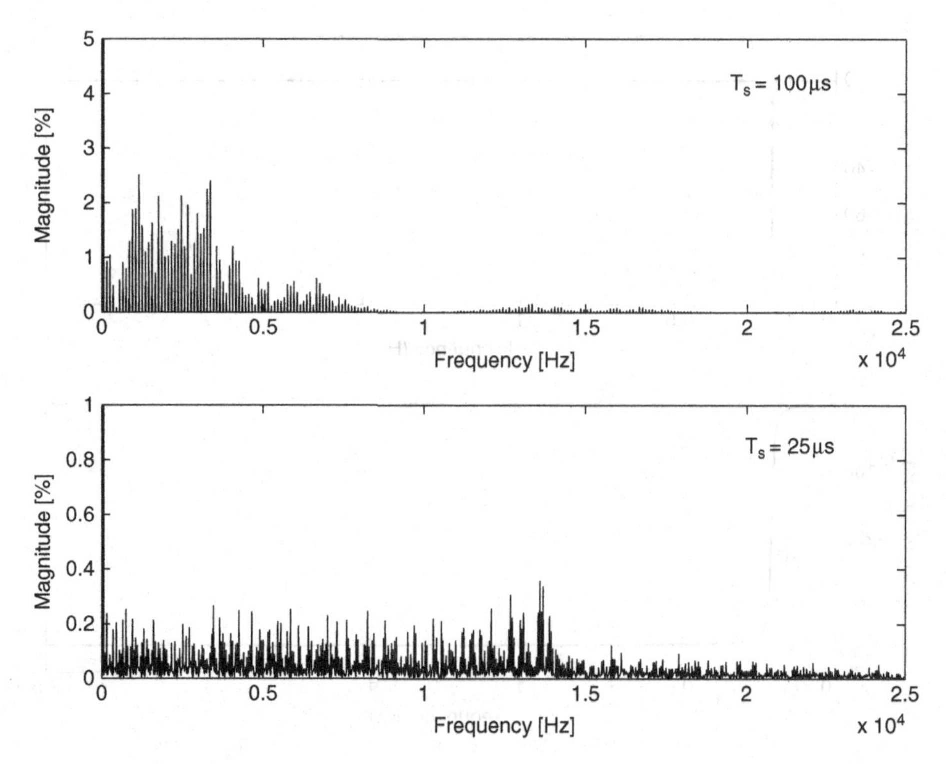

Figure 10.9 Load current spectrum for predictive current control of a three-phase inverter operating at different sampling frequencies (Cortes *et al.*, 2008 © IEEE)

In some applications, such a spread spectrum is not desirable because it can produce oscillations and make the design of passive filters difficult. In order to overcome these problems it is possible to shape the spectrum by introducing a filter in the cost function. In this way, different frequencies have different weights in the cost function allowing control of the harmonic content in the controlled variables.

The modified cost function for spectrum shaping is defined as

$$g = \left|\left|F(x^* - x^p)\right|\right| \tag{10.24}$$

where F is a digital filter.

The discrete-time filter F can be implemented by the following transfer function:

$$F(z) = \frac{z^0 + b_1 z^{-1} + \cdots + b_n z^{-n}}{a_0 z^0 + a_1 z^{-1} + \cdots + a_n z^{-n}} \tag{10.25}$$

where n is the order of the filter.

In the design of the F filter it is possible to shape the spectrum of the controlled variable. Since the frequency response of the filter defines the weight of each frequency in the cost function, the resulting spectrum will be similar to the inverse of the frequency response of the filter. For example, if a band stop filter with a center frequency f_o is used, the harmonic content will be concentrated around f_o, and if a low-pass filter is

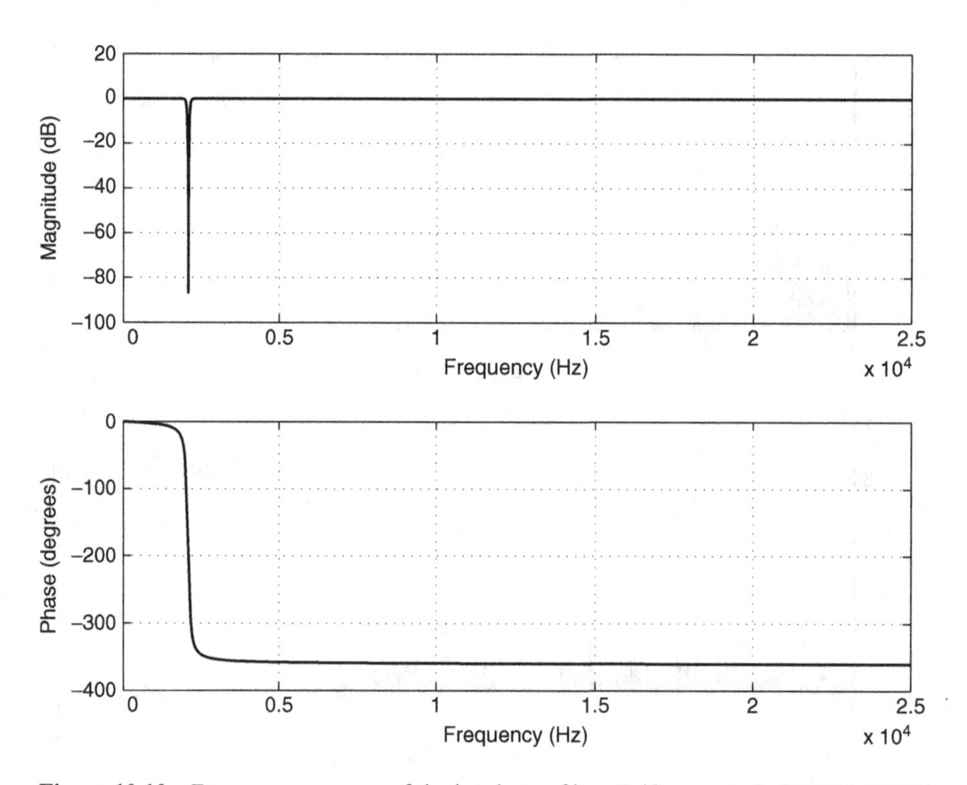

Figure 10.10 Frequency response of the band stop filter F (Cortes *et al.*, 2008 © IEEE)

used, the spectral content will be concentrated in the high-frequency range over the cut off frequency of the filter.

In order to obtain a similar spectrum to the one obtained using PWM, it is necessary to concentrate the spectral content in a narrow frequency range using a band stop filter. For current control in a three-phase inverter, the following cost function can be considered [8]:

$$g = \left|F(i_\alpha^* - i_\alpha^p)\right| + \left|F(i_\beta^* - i_\beta^p)\right| \tag{10.26}$$

Here the filter F is defined as a band stop filter with the frequency response shown in Figure 10.10. Results using this filter with a center frequency of 2 kHz are shown in Figure 10.11. It can be seen that the load current spectrum is concentrated around the defined frequency, avoiding the presence of harmonic content over a wide range of frequencies.

A different approach for including spectral content information in the cost function is to use the discrete Fourier transform (DFT) in order to control the value of individual harmonics of the controlled variables. This idea was proposed in [9] for the implementation of closed loop predictive harmonic elimination for multilevel converters.

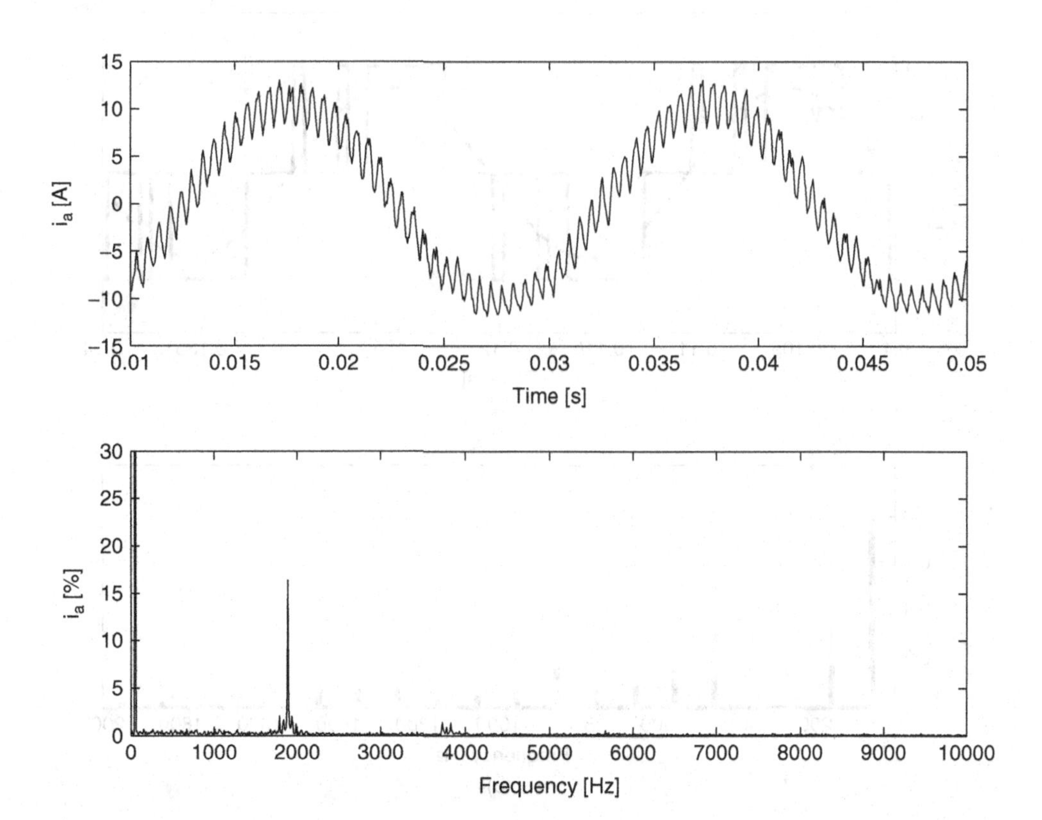

Figure 10.11 Load current and spectrum for predictive control using a band stop filter F with the center frequency at 2 kHz (Cortes *et al.*, 2008 © IEEE)

These kinds of cost functions have interesting applications in high-power systems. As an example, a high-power NPC inverter is considered, where the control objectives are very low switching frequency, good tracking of the fundamental output voltage, and elimination of several low-frequency harmonics. The cost function for this predictive harmonic elimination application is expressed as

$$g = \mathrm{DFT}_{f1}\left\{\left|v^* - v^P\right|\right\} + \lambda_f \sum_i^N \mathrm{DFT}_{fi}\left\{\left|v^* - v^P\right|\right\} + \lambda_{sw} n \tag{10.27}$$

where the first term is the DFT of the voltage error at fundamental frequency and allows tracking of the fundamental of the output voltage, the second term allows elimination of different harmonics fi, with $i = 0, 2, 3, 4, \ldots, N$, and the last term allows reduction of the switching frequency as explained in previous sections.

Results using this control strategy are shown in Figure 10.12 for the elimination of harmonics 0, 2, 4, 6, 7, 8, and 10, operating at a switching frequency of 300 Hz.

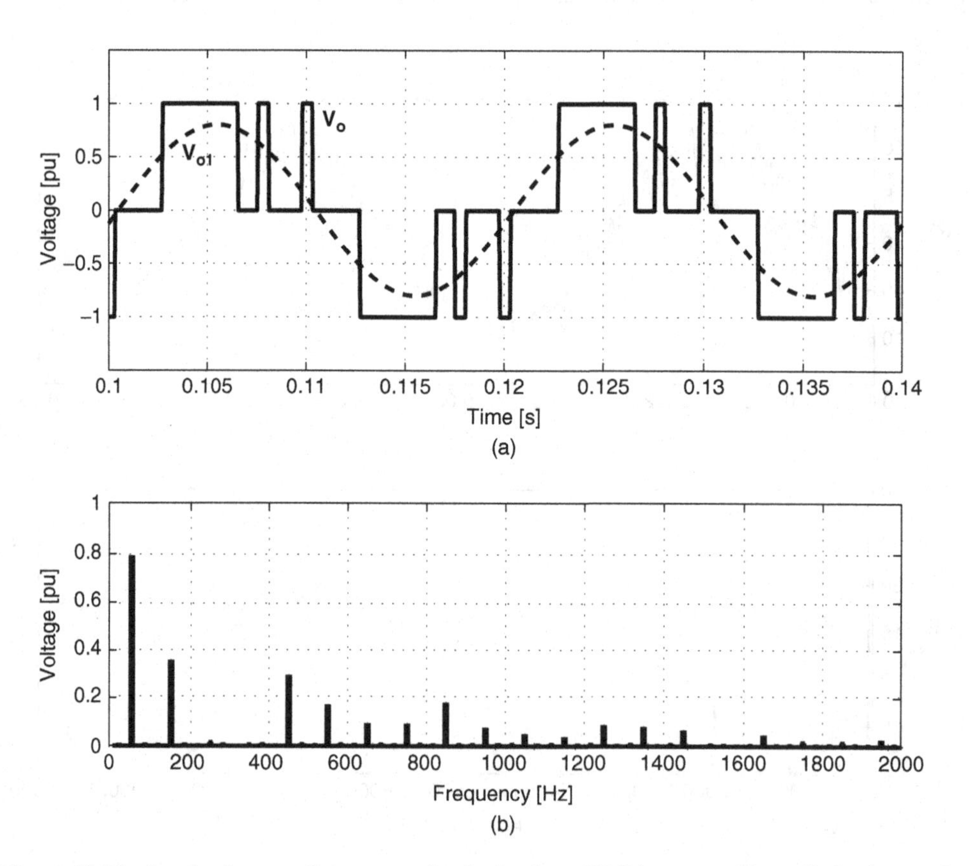

Figure 10.12 Load voltage and spectrum of a single-phase NPC inverter with predictive harmonic elimination (Kouro *et al.*, 2009 © IEEE)

10.6 Summary

This chapter presents different types of terms that can be included in the cost function. Some examples are included to illustrate each type. The cost function terms can be classified as reference following, actuation constraints, hard constraints, and spectral content terms. These terms constitute the building blocks that allow the system requirements to be expressed in the form of a cost function.

References

[1] J. Rodríguez, J. Pontt, C. Silva *et al.*, "Predictive current control of a voltage source inverter," *IEEE Transactions on Industrial Electronics*, vol. 54, no. 1, pp. 495–503, February 2007.

[2] P. Cortés, J. Rodríguez, P. Antoniewicz, and M. Kazmierkowski, "Direct power control of an AFE using predictive control," *IEEE Transactions on Power Electronics*, vol. 23, no. 5, pp. 2516–2523, September 2008.

[3] R. Vargas, J. Rodríguez, U. Ammann, and P. W. Wheeler, "Predictive current control of an induction machine fed by a matrix converter with reactive power control," *IEEE Transactions on Industrial Electronics*, vol. 55, no. 12, pp. 4362–4371, December 2008.

[4] R. Vargas, P. Cortés, U. Ammann, J. Rodríguez, and J. Pontt, "Predictive control of a three-phase neutral-point-clamped inverter," *IEEE Transactions on Industrial Electronics*, vol. 54, no. 5, pp. 2697–2705, October 2007.

[5] F. Schafmeister, C. Rytz, and J. Kolar, "Analytical calculation of the conduction and switching losses of the conventional matrix converter and the (very) sparse matrix converter," in Conference Record of Applied Power Electronics Conference and Exposition, APEC 2005, pp. 875–881, March 2005.

[6] R. Vargas, U. Ammann, and J. Rodríguez, "Predictive approach to increase efficiency and reduce switching losses on matrix converters," *IEEE Transactions on Power Electronics*, vol. 24, no. 4, pp. 894–902, April 2009.

[7] H. Miranda, P. Cortés, J. I. Yuz, and J. Rodríguez, "Predictive torque control of induction machines based on state-space models," *IEEE Transactions on Industrial Electronics*, vol. 56, no. 6, pp. 1916–1924, June 2009.

[8] P. Cortés, J. Rodríguez, D. E. Quevedo, and C. Silva, "Predictive current control strategy with imposed load current spectrum," *IEEE Transactions on Power Electronics*, vol. 23, no. 2, pp. 612–618, March 2008.

[9] S. Kouro, B. La Rocca, P. Cortés *et al.*, "Predictive control based selective harmonic elimination with low switching frequency for multilevel converters," in IEEE Energy Conversion Congress and Exposition, 2009, ECCE, pp. 3130–3136, September 2009.

11

Weighting Factor Design

11.1 Introduction

One of the major advantages of model predictive control (MPC) is that several control targets, variables, and constraints can be included in a single cost function and simultaneously controlled. In this way typical variables such as current, voltage, torque, or flux can be controlled while achieving additional control requirements like switching frequency reduction, common-mode voltage reduction, and reactive power control, to name just a few. This can be accomplished simply by introducing the additional control targets in the cost function to be evaluated. However, the combination of two or more variables in a single cost function is not a straightforward task when they are of a different nature (different units and different orders of magnitude in value). Each additional term in the cost function has a specific weighting factor, which is used to tune the importance or cost of that term in relation to the other control targets. These parameters have to be properly designed in order to achieve the desired performance. Unfortunately, there are no analytical or numerical methods or control design theories to adjust these parameters, and currently they are determined based on empirical procedures. Although this challenge has not prevented MPC from being applied successfully to several power converters, it is highly desirable to establish a procedure or define some basic guidelines to reduce the uncertainty and improve the effectiveness of the tuning stage.

This chapter presents a first approach to address this challenge. First, some representative examples of MPC cost functions are classified according to the nature of their terms, in order to group types of weighting factors that could be tuned similarly. Then a set of simple guidelines is analyzed and tested to evaluate the evolution of system performance in relation to changes in the weighting factors. Several converter and drive control applications will be studied to cover a wide variety of cost functions and weighting factors. In addition, results for three different weighting factors are presented to compare results and validate the methodology.

Predictive Control of Power Converters and Electrical Drives, First Edition. Jose Rodriguez and Patricio Cortes.

11.2 Cost Function Classification

Although the cost function's main objective is to keep track of a particular variable and control the system, it is not limited to this, as explained in the previous chapter. In fact, one of the main advantages of MPC is that the cost function admits any necessary term that could represent a prediction for another system variable, system constraint, or system requirement. Since these terms most likely can be of a different physical nature (current, voltage, reactive power, switching losses, torque, flux, etc.) their units and magnitudes can also be very different. This issue has been commonly dealt with in MPC by including weighting coefficients or weighting factors λ, for each term of the cost function

$$g = \lambda_x||x^* - x^p|| + \lambda_y||y^* - y^p|| + \cdots + \lambda_z||z^* - z^p|| \tag{11.1}$$

Depending on the nature of the different terms involved in the formulation of the cost function, they can be classified in different groups. This classification is necessary in order to facilitate the definition of a weighting factor adjustment procedure that could be applied to similar types of cost functions or terms.

11.2.1 Cost Functions without Weighting Factors

In this kind of cost function, only one, or the components of one variable, are controlled. This is the simplest case, and since only one type of variable is controlled, no weighting factors are necessary. Some representative examples of this type of cost function are obtained for: predictive current control of a voltage source inverter [1]; predictive power control of an active front-end (AFE) rectifier [2]; predictive voltage control of an uninterruptible power supply (UPS) system [3]; predictive current control with imposed switching frequency [4]; and predictive current control in multi-phase inverters [5–7], among others. The corresponding cost functions are summarized in Table 11.1.

Note that all the terms in a cost function are composed of variables of the same nature (same unit and order of magnitude). Moreover, some of them are a decomposition of a single vector into two or more components. Therefore, no weighting factors or their corresponding tuning processes are necessary.

11.2.2 Cost Functions with Secondary Terms

Some systems have a primary goal or a more important control objective that must be achieved in order to provide a proper system behavior, and additional secondary constraints or requirements that should also be accomplished to improve system performance, efficiency, or power quality. In this case the cost function contains primary and secondary terms, where the importance of the secondary terms can vary over a wide range, depending on the application and its specific needs. Some examples are: predictive current control with reduction of the switching frequency to improve efficiency [8]; predictive current control with reduction of common-mode voltage to prevent motor damage [9]; and predictive current control with reactive power reduction to improve power quality [10, 11]. The corresponding cost functions are listed in Table 11.2.

The importance of the second term (i.e., how much the switching frequency, the common-mode voltage, or the reactive power is reduced), will depend on the specific

Table 11.1 Cost functions without weighting factors

Application	Cost function
Current control of a VSI	$\lvert i_\alpha^* - i_\alpha^p\rvert + \lvert i_\beta^* - i_\beta^p\rvert$
Power control of an AFE rectifier	$\lvert Q^p\rvert + \lvert P^* - P^p\rvert$
Voltage control of a UPS	$(v_{c\alpha}^* - v_{c\alpha}^p)^2 + (v_{c\beta}^* - v_{c\beta}^p)^2$
Imposed switching frequency in a VSI	$\lvert F(i_\alpha^* - i_\alpha^p)\rvert + \lvert F(i_\beta^* - i_\beta^p)\rvert$
Current control of a multi-phase VSI	$\lvert i_\alpha^* - i_\alpha^p\rvert + \lvert i_\beta^* - i_\beta^p\rvert + \lvert i_x^* - i_x^p\rvert + \lvert i_y^* - i_y^p\rvert$

Table 11.2 Cost functions with secondary terms

Application	Cost function
Switching frequency reduction	$\lvert i_\alpha^* - i_\alpha^p\rvert + \lvert i_\beta^* - i_\beta^p\rvert + \lambda_{sw} n_{sw}^p$
Common-mode voltage reduction	$\lvert i_\alpha^* - i_\alpha^p\rvert + \lvert i_\beta^* - i_\beta^p\rvert + \lambda_{cm} \lvert V_{cm}^p\rvert$
Reactive power reduction	$\lvert i_\alpha^* - i_\alpha^p\rvert + \lvert i_\beta^* - i_\beta^p\rvert + \lambda_Q \lvert Q^p\rvert$

Table 11.3 Cost functions with equally important terms

Application	Cost function
Torque and flux control	$1/T_{en}^2 (T_e^* - T_e^p)^2 + \lambda_\psi / \psi_{sn}^2 (\lvert\psi_s\rvert^* - \lvert\psi_s^p\rvert)^2$
Capacitor voltage balance	$1/i_{sn} \left[\lvert i_\alpha^* - i_\alpha^p\rvert + \lvert i_\beta^* - i_\beta^p\rvert\right] + \lambda_{\Delta V} / V_{cn} \lvert\Delta V_c^p\rvert$

needs of the application and will impose a trade-off with the primary control objective, in this case current control. Note that in each cost function a weighting factor λ is included with the corresponding secondary term. Hence, solving the trade-off can be seen as a weighting factor adjustment to the cost function.

11.2.3 Cost Functions with Equally Important Terms

Unlike the previous case, there are systems in which several variables need to be controlled simultaneously with equal importance in order to control the system. Here the cost function can include several terms with equal importance, and it is the job of the weighting factors to compensate the difference in nature of the variables. Such is the case for the torque and flux control of an induction machine, where both variables need to be controlled accurately to achieve proper system performance [12, 13]. Another example is the current control of a neutral-point clamped (NPC) inverter, in which the DC link capacitor voltage balance is essential in order to reduce voltage distortion and avoid system damage (exceed the permitted voltage level of the capacitors, otherwise overrated capacitors should be used) [8]. Both cost functions are included in Table 11.3. Note that there are two additional terms in each cost function used to normalize the quantities in relation to their nominal values (denoted by subscript n); the reason for this will be explained later.

11.3 Weighting Factors Adjustment

The weighting factor tuning procedure will vary depending on which types of terms are present in the cost function, as classified in the previous section.

11.3.1 For Cost Functions with Secondary Terms

This is the easiest case for weighting factor adjustment, since the system can be first controlled using only the primary control objective or term. This can be very simply achieved by neglecting the secondary terms forcing the weighting factor to zero ($\lambda = 0$). Hence the first step of the procedure is to convert the cost function with secondary terms into a cost function without weighting factors. This will set the starting point for the measurement of the behavior of the primary variable.

The second step is to establish measurements or figures of merit that will be used to evaluate the performance achieved by the weighting factor. For all the examples given in Table 11.2 a straightforward quantity should be one related to the primary variable, which is current error. Several error measures for current can be defined, such as the root mean square (RMS) value of the error at steady state, or the total harmonic distortion (THD). At least one additional measure is necessary to establish the trade-off with the secondary term. For the three cost functions of Table 11.2 the corresponding measures that can be selected are: the device average switching frequency f_{sw}, the RMS common-mode voltage, and the steady state input reactive power.

Once the measures are defined, the procedure is as follows. Evaluate the system behavior with simulations starting with $\lambda = 0$ and increase the value gradually. Record the corresponding measures for each value of λ. Stop the increments of λ once the measured value for the secondary term has reached the desired value for the specific application, or keep increasing λ until the primary variable is not properly controlled. Then plot the results and select a value of λ that fulfills the system requirements for both variables. This procedure can be programmed by automating and repeating the simulation, introducing an increment in the weighting factor after each simulation.

In order to reduce the n umber of simulations required to find a proper value for the weighting factor, a branch and bound algorithm can be used. For this approach, first select a couple of initial values for the weighting factor λ, usually with different orders of magnitude to cover a very wide range, for example, $\lambda = 0$, 0.1, 1, and 10. A qualitative example of this algorithm is illustrated in Figure 11.1. Then simulate these weighting factors and obtain the measures for both terms, M^1 and M^2, for the primary and secondary terms respectively. Then compare these results to the desired maximum errors admitted by the application and fit them into an interval of two weighting factors ($0.1 \leq \lambda \leq 1$ in the example). Then compute the measures for the λ in half of the new interval ($\lambda = 0.5$ in the example) and continue until a suitable λ is achieved. Note in Figure 11.1 that each solid line corresponds to a simulation and dashed lines correspond to values already simulated. This method reduces the number of simulations necessary to obtain a working weighting factor.

The qualitative example of Figure 11.1 can be matched with the results for the common-mode reduction case shown below in Figure 11.3a. Note that with only six simulations

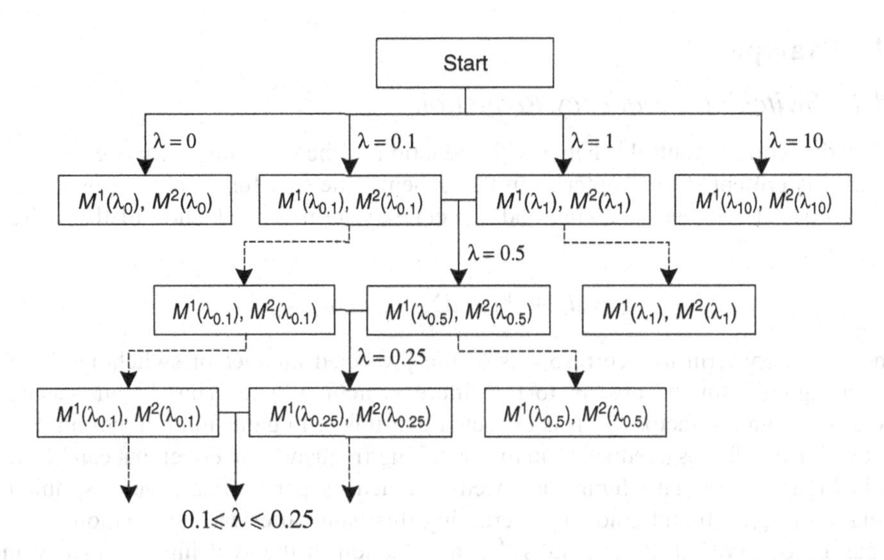

Figure 11.1 Branch and bound algorithm to reduce the number of simulations required to obtain suitable weighting factors (Cortes *et al.*, 2009 © IEEE)

the search for λ_{cm} would have narrowed to an interval $0.1 \leq \lambda_{cm} \leq 0.25$ where any λ_{cm} would work properly.

11.3.2 For Cost Functions with Equally Important Terms

For cost functions like those listed in Table 11.3, a different procedure needs to be considered since λ is not allowed to be zero.

As a first step the different nature of the variables has to be considered. For example, when controlling torque and flux in an adjustable speed drive application with a rated torque and flux of 25 N m and 1 Wb respectively, the torque error can have different orders of magnitudes making both variables not equally important in the cost function and affecting the system performance. Thus the first step is to normalize the cost function. Once normalized, all the terms will be equally important and now $\lambda = 1$ can be considered as the starting point. Usually a suitable λ is located close to 1. Note that the cost functions in Table 11.3 have already included this normalization (nominal values are denoted by subscript n).

The second step is the same as in the previous procedure, that is, measurements or figures of merit have to be defined in order to evaluate the performance achieved by each weighting factor value.

The last step is to perform the branch and bound algorithm of Figure 11.1 for a couple of starting points. Naturally $\lambda = 1$ has to be considered, but $\lambda = 0$ is not a possible alternative. When a small interval of weighting factors has been reached, meaning, by small interval, that there are no big differences in the measured values obtained with the upper and lower bounds of the interval, then the weighting factor has been obtained.

11.4 Examples

11.4.1 Switching Frequency Reduction

A predictive current control scheme with reduction of the switching frequency for a NPC inverter was presented in Chapter 5. In this scheme, the cost function presents a primary term for current reference tracking and a secondary term for reduction of the switching frequency:

$$g = |i_\alpha^* - i_\alpha^{\mathbf{p}}| + |i_\beta^* - i_\beta^{\mathbf{p}}| + \lambda_{sw} n_{sw}^p \tag{11.2}$$

The secondary term n_{sw}^p corresponds to the predicted number of switchings involved when changing from the present to the future switching state. Thus by increasing the associated weighting factor λ_{sw} it is expected that this term gains more importance in the cost function and forces a reduction in the switching frequency, an effect that can be clearly seen in Figure 11.2a. This figure has been obtained by performing several simulations, starting with $\lambda_{sw} = 0$ and gradually increasing this value after each simulation.

It can be observed in these results that a reduction in the switching frequency introduces higher distortion, affecting the quality of the load current. This trade-off is very clear in Figure 11.2a since the curves representing each measure have opposite evolutions for the different values of λ_{sw}. A suitable selection of λ_{sw} would be any value in $0.04 \leq \lambda_{sw} \leq 0.06$ since the current error is still below 10% of the rated current (15 A in this example) and a reduction from 1000 to 500 Hz is achieved for the average device switching frequency. Finally $\lambda_{sw} = 0.05$ has been selected. In this application it is also possible to select λ_{sw} in order to achieve a given switching frequency.

Figure 11.2b shows comparative results for the system working with three different values of λ_{sw}, one of them the selected value. Note how the load current presents higher distortion for the larger value of λ_{sw} due to the strong reduction in the number of commutations. On the other hand, for $\lambda_{sw} = 0$ the current control works at its best, but at the expense of higher switching losses. Since the NPC is intended for medium-voltage, high-power applications where losses become important, the selection of $\lambda_{sw} = 0.05$ merges efficiency with performance. A comparison of this predictive method and a traditional PWM-based controller can be found in [8].

11.4.2 Common-Mode Voltage Reduction

The guidelines presented in this chapter have been used to tune the weighting factor of the second equation in Table 11.2. This example corresponds to the predictive current control of a matrix converter with reduction of the common-mode voltage [9], using the following cost function:

$$g = |i_\alpha^* - i_\alpha^{\mathbf{p}}| + |i_\beta^* - i_\beta^{\mathbf{p}}| + \lambda_{cm} |V_{cm}^p| \tag{11.3}$$

where the additional term V_{cm}^p is the predicted common-mode voltage for the different switching states and will be considered a secondary objective of the control; its effect is adjusted by properly tuning the weighting factor λ_{cm}.

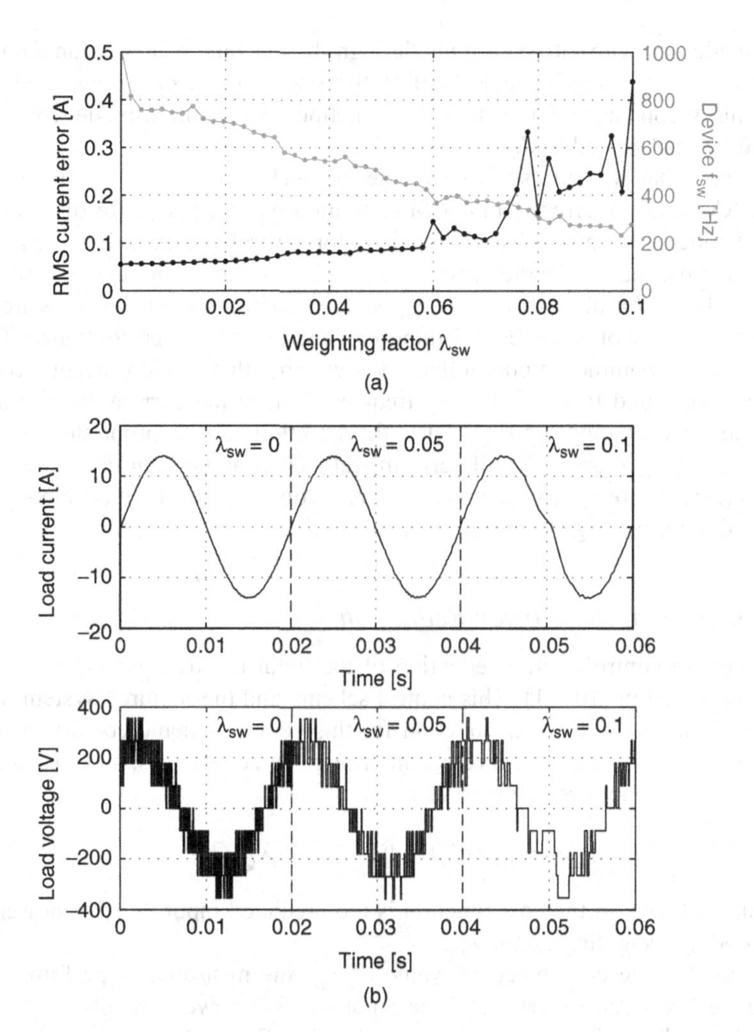

Figure 11.2 (a) Weighting factor influence on the current error and the device average switching frequency f_{sw}. (b) Results comparison for different weighting factors (load current and load voltage) (Cortes *et al.*, 2009 © IEEE)

The predictive current control strategy for the matrix converter is explained in Chapter 7. The common-mode voltage is defined as

$$V_{cm} = \frac{v_{aN} + v_{bN} + v_{cN}}{3} \tag{11.4}$$

where the output voltages v_{aN}, v_{bN}, and v_{cN} are calculated as a function of the voltages at the input of the matrix converter and the converter switching states. Common-mode voltages cause overvoltage stress in the winding insulation of the electrical machine fed by the power converter, producing deterioration and reducing the lifetime of the machine. In addition, capacitive currents through the machine bearings can damage them, and electromagnetic interference can affect the operation of electronic equipment. By including this secondary term in the cost function the common-mode voltage and its negative effects can be reduced.

The measures that will be used to evaluate the performance of the different values of λ_{cm} are the RMS current error and the RMS common-mode voltage. Results showing both measures obtained from a series of simulations for several values of λ_{cm} are presented in Figure 11.3a. Note that, as in the previous example, similar evolutions of both measures are obtained, that is, for higher values of λ_{cm} smaller common-mode voltages are obtained, while the current control becomes less important and loses some performance. The results also show that the common-mode voltage is a variable that is more decoupled from the load current, compared to the switching frequency, since the current error remains very low throughout a wide range of λ_{cm}. Hence the selection of an appropriate value is easier, and values of $0.05 \leq \lambda_{cm} \leq 0.5$ will perform well. This can be seen for the results shown in Figure 11.3b, where clearly a noticeable reduction in the common-mode voltage is achieved without affecting the current control.

11.4.3 Input Reactive Power Reduction

Predictive current control with a reduction of the input reactive power in a matrix converter was proposed in [10, 11]. This control scheme and the required system models are explained in Chapter 7. The cost function for this control scheme consists of a primary term for output current control, expressed in orthogonal coordinates, and a secondary term for reduction of the input reactive power:

$$g = |i_\alpha^* - i_\alpha^p| + |i_\beta^* - i_\beta^p| + \lambda_Q |Q^p| \tag{11.5}$$

The additional term in the cost function is the predicted input reactive power Q^p with its corresponding weighting factor λ_Q.

In order to evaluate and select the value of λ_Q, the measures of performance of the system are the RMS current error and the input reactive power magnitude Q.

The results of the tuning procedure are depicted in Figure 11.4a. Since this cost function belongs to the same classification as the previous two examples, it is expected to present similar measurement evolution with increasing λ_Q. As in the previous case, the input reactive power seems to be very decoupled from the load current, hence the current error remains very low for a wide range of λ_Q. It becomes easy to obtain a suitable value by considering $0.05 \leq \lambda_Q \leq 0.25$. This can be corroborated by the results given in Figure 11.4b, showing an important reduction of the input reactive power for $\lambda_Q = 0.05$.

11.4.4 Torque and Flux Control

A good example of a cost function with equally important terms is the predictive torque and flux control for an induction machine, which is explained in Chapter 8. Here, the objective of the control algorithm is to simultaneously control the electrical torque T_e and the magnitude of the stator flux ψ_s. This objective can be expressed as a cost function with

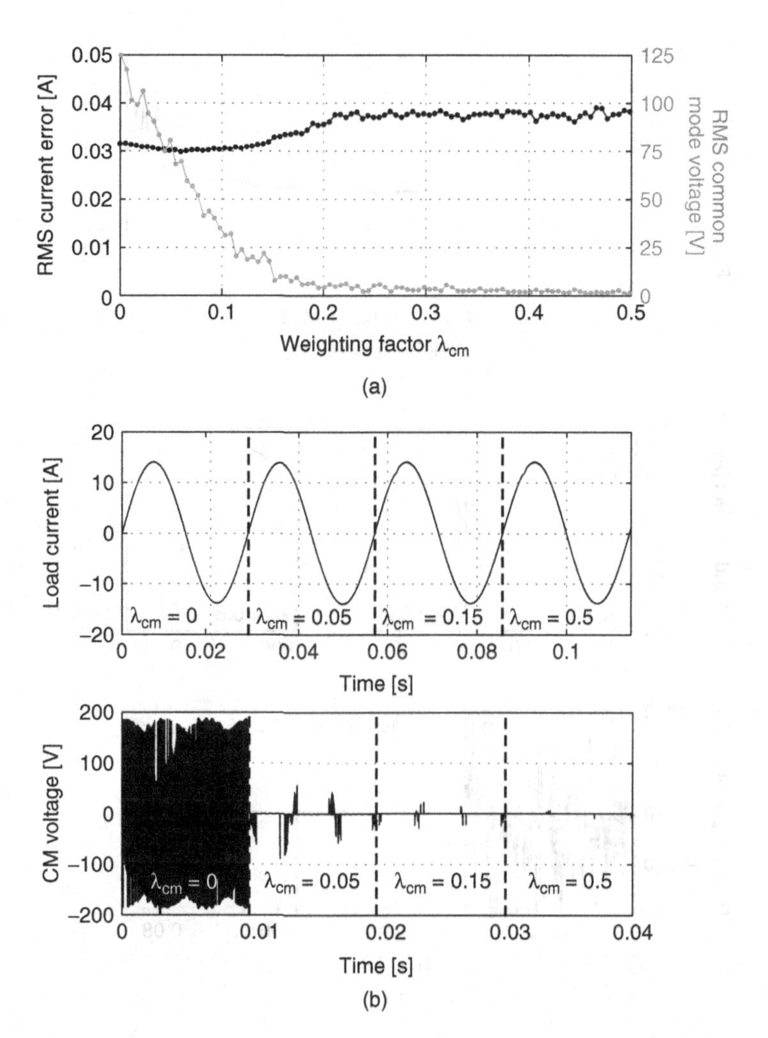

Figure 11.3 (a) Weighting factor influence on the current error and the common-mode voltage. (b) Results comparison for different weighting factors (load current and common-mode voltage) (Cortes *et al.*, 2009 © IEEE)

two terms, torque error and flux error, and the weighting factor must handle the difference in magnitude and units between these two terms, as proposed in [12]. A different approach consists of using a normalized cost function where each term is divided by its rated value, resulting in

$$g = \frac{(T_e^* - T_e^p)^2}{T_{er}^2} + \lambda_\psi \frac{(|\boldsymbol{\psi}_s|^* - |\boldsymbol{\psi}_s|^p)^2}{|\boldsymbol{\psi}_{sr}|^2} \tag{11.6}$$

Using this cost function, the same importance is given to both terms using $\lambda_\psi = 1$, as proposed in [13]. However, the optimal value can be different, depending on the defined criteria for optimal operation.

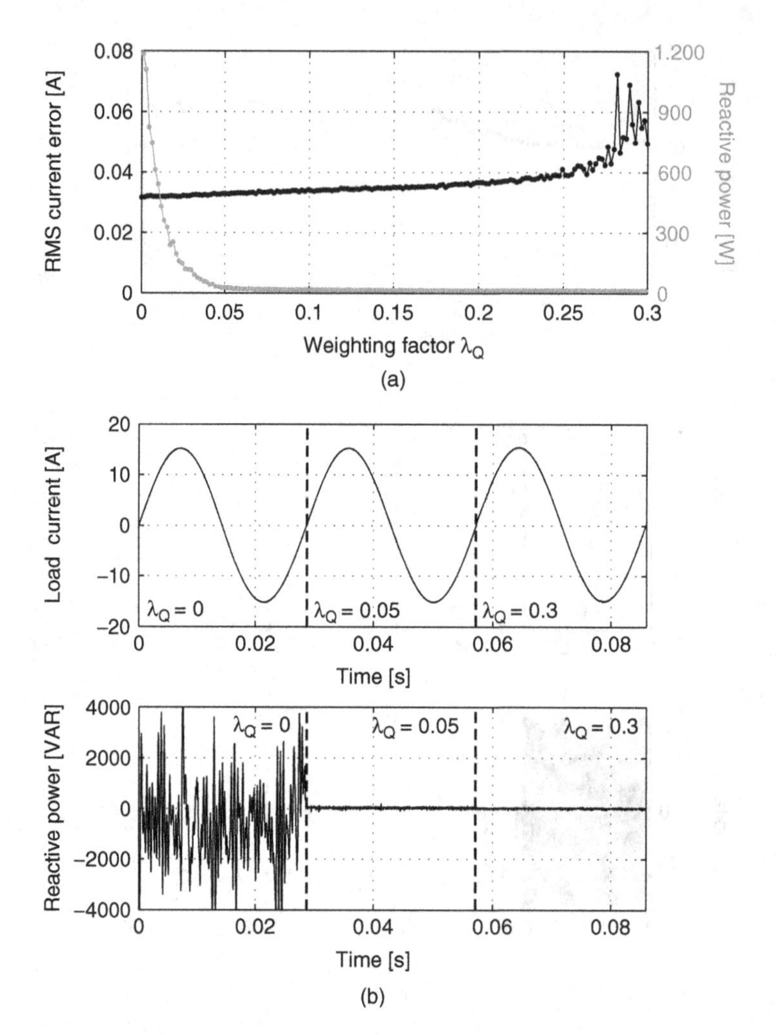

Figure 11.4 (a) Weighting factor influence on the current error and the input reactive power. (b) Results comparison for different weighting factors (load current and input reactive power) (Cortes *et al.*, 2009 © IEEE)

In order to evaluate the performance of the control for different values of the weighting factor λ_ψ, the RMS torque error and RMS stator flux magnitude error are defined as measures of performance.

A branch and bound algorithm starting with $\lambda_\psi = 0.01, 0.1, 1, 10$, and 100 first gave the interval $0.1 \leq \lambda_\psi \leq 1$, and then $0.5 \leq \lambda_\psi \leq 1$, with very small differences. Finally $\lambda_\psi = 0.85$ was chosen. Note that the obtained optimal value is very close to the initial value of $\lambda_\psi = 1$.

Figure 11.5a shows extensive results considering many more values of λ_ψ (note that the values are represented in $\log_{10}$ scale), to show that the branch and bound method really has found a suitable solution.

Results for different values of λ_ψ, including $\lambda_\psi = 0.85$, are given in Figure 11.5b to show the performance achieved by the predictive torque and flux control. Note that

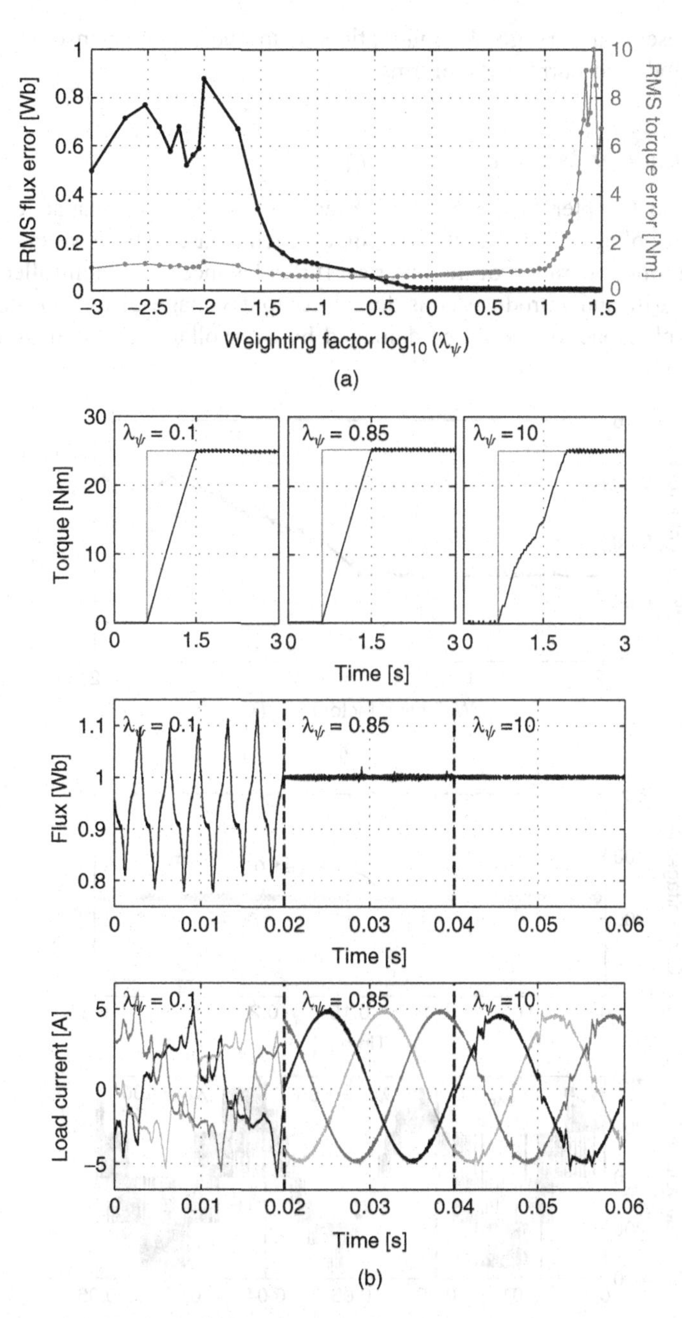

Figure 11.5 (a) Weighting factor influence on the flux and torque errors. (b) Results comparison for different weighting factors (torque step response, flux magnitude at steady state, and load currents) (Cortes *et al*., 2009 © IEEE)

$\lambda_{\psi} = 0.85$ presents a very good combination of torque step response and steady state, flux control, and load current waveforms.

11.4.5 Capacitor Voltage Balancing

As presented in Chapter 5, the NPC inverter has two DC link capacitors in order to generate three voltage levels at the output of each phase. These voltages need to be balanced for proper operation of the inverter. If this balance is not controlled, the DC link voltages will drift and introduce considerable output voltage distortion, not to mention that the DC link capacitors could be damaged by overvoltage, unless they are overrated.

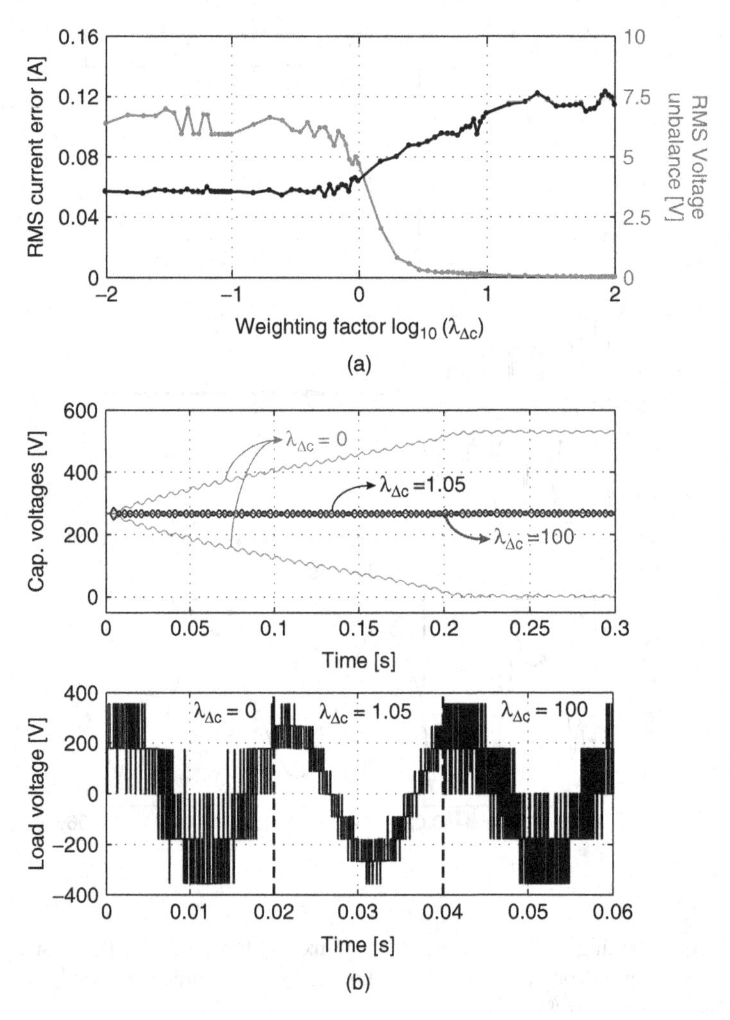

Figure 11.6 (a) Weighting factor influence on the current error and the DC link capacitor unbalance. (b) Results comparison for different weighting factors (load current and DC link capacitor voltages, dynamic behavior) (Cortes *et al*., 2009 © IEEE)

This control requirement can be considered in the predictive current control scheme by introducing an additional term in the cost function. The resulting cost function is expressed as

$$g = \frac{1}{i_{sn}}\left[|i_\alpha^* - i_\alpha^p| + |i_\beta^* - i_\beta^p|\right] + \frac{\lambda_{\Delta C}}{V_{cn}}|\Delta V_c^p| \tag{11.7}$$

where ΔV_c^p is the predicted voltage unbalance and $\lambda_{\Delta C}$ is the weighting factor to be adjusted. Note that the cost function terms have been normalized considering the values of rated current i_{sn} and rated DC link capacitor voltage V_{cn}, as indicated in the first step of the adjustment procedure.

The measures that will be used to evaluate the weighting factor $\lambda_{\Delta C}$ are the RMS values of the current error and of the voltage unbalance.

A branch and bound algorithm starting with $\lambda_{\Delta C} = 10^{-2}$, 10^{-1}, 1, 10^1, and 10^2 first gave the interval $1 \leq \lambda_{\Delta C} \leq 10$; after this first evaluation two additional iterations were necessary until very small differences were obtained between the extremes of the interval. Finally $\lambda_{\Delta C} = 1.05$ was chosen. Figure 11.6a shows extensive results considering many more values (note that the values are represented in $\log_{10}$ scale), to show that the branch and bound method really has found a suitable solution.

Results for different $\lambda_{\Delta C}$, including $\lambda_{\Delta C} = 1.05$, are given in Figure 11.6b to show the performance achieved by the MPC. Note that for $\lambda_{\Delta C} = 0$, which normally is not allowed since it does not control the unbalance producing the maximum drift of the DC link capacitors, the load voltage only presents five different voltage levels, while nine levels should appear in the load phase-to-neutral voltage (since the NPC has three levels in the converter phase-to-neutral voltage). Only five appear since the NPC is not generating three output levels but only two, due to the voltage drift of its capacitors. On the other hand, $\lambda_{\Delta C} = 100$ controls the voltage unbalance very accurately; it even makes voltage unbalance so important in the cost function that it avoids generation of those switching states that produce unbalance, eliminating voltage levels at the output and increases the switching frequency, as can be seen in the load voltage of Figure 11.6b. Finally, the selected weighting factor value $\lambda_{\Delta C} = 1.05$ presents the nine load voltage levels, controls the load current, and keeps the capacitor voltages balanced.

11.5 Summary

This chapter presents an empirical procedure for adjusting the weighting factors of the cost function in a predictive control scheme.

Three types of cost function have been identified and a procedure for each type is presented. In this way, the first stage of weighting factor adjustment is to identify the type of cost function that will be used in the control.

At least two different figures of merit or system parameters have to be considered, depending on the application, to settle the trade-off present in the design choice of the weighting factors.

For cost functions with a primary control objective and secondary terms, the starting point is $\lambda = 0$. Then increments in the value of λ are tested until the desired behavior is obtained (branch and bound can also be used).

For cost functions with equally important terms, the cost function must be normalized, and then the weighting factor must be set to $\lambda = 1$. With this value the system will be controlled and, for fine tuning, branch and bound can be used or slight variations of λ around 1 can be tested.

Examples of the presented procedures are provided for adjusting the weighting factors of several predictive control applications.

References

[1] J. Rodríguez, J. Pontt, C. Silva *et al.*, "Predictive current control of a voltage source inverter," *IEEE Transactions on Industrial Electronics*, vol. 54, no. 1, pp. 495–503, February 2007.
[2] P. Cortés, J. Rodríguez, P. Antoniewicz, and M. Kazmierkowski, "Direct power control of an AFE using predictive control," *IEEE Transactions on Power Electronics*, vol. 23, no. 5, pp. 2516–2523, September 2008.
[3] P. Cortés, G. Ortiz, J. I. Yuz *et al.*, "Model predictive control of an inverter with output LC filter for UPS applications," *IEEE Transactions on Industrial Electronics*, vol. 56, no. 6, pp. 1875–1883, June 2009.
[4] P. Cortés, J. Rodríguez, D. E. Quevedo, and C. Silva, "Predictive current control strategy with imposed load current spectrum," *IEEE Transactions on Power Electronics*, vol. 23, no. 2, pp. 612–618, March 2008.
[5] F. Barrero, M. Arahal, R. Gregor, S. Toral, and M. Duran, "A proof of concept study of predictive current control for VSI-driven asymmetrical dual three-phase AC machines," *IEEE Transactions on Industrial Electronics*, vol. 56, no. 6, pp. 1937–1954, June 2009.
[6] P. Cortés, L. Vattuone, J. Rodríguez, and M. Duran, "A method of predictive current control with reduced number of calculations for five-phase voltage source inverters," in 35th IEEE Annual Conference on Electronics IECON '09, November 2009, pp. 53–58.
[7] M. J. Duran, J. Prieto, F. Barrero, and S. Toral, "Predictive current control of dual three-phase drives using restrained search techniques," *IEEE Transactions on Industrial Electronics*, vol. 58, no. 8, pp. 3253–3263, August 2011.
[8] R. Vargas, P. Cortés, U. Ammann, J. Rodríguez, and J. Pontt, "Predictive control of a three-phase neutral-point-clamped inverter," *IEEE Transactions on Industrial Electronics*, vol. 54, no. 5, pp. 2697–2705, October 2007.
[9] R. Vargas, U. Ammann, J. Rodríguez, and J. Pontt, "Predictive strategy to control common-mode voltage in loads fed by matrix converters," *IEEE Transactions on Industrial Electronics*, vol. 55, no. 12, pp. 4372–4380, December 2008.
[10] S. Muller, U. Ammann, and S. Rees, "New time-discrete modulation scheme for matrix converters," *IEEE Transactions on Industrial Electronics*, vol. 52, no. 6, pp. 1607–1615, December 2005.
[11] R. Vargas, J. Rodríguez, U. Ammann, and P. W. Wheeler, "Predictive current control of an induction machine fed by a matrix converter with reactive power control," *IEEE Transactions on Industrial Electronics*, vol. 55, no. 12, pp. 4362–4371, December 2008.
[12] J. Rodríguez, J. Pontt, C. Silva *et al.*, "Predictive direct torque control of an induction machine," in EPE-PEMC 2004 (Power Electronics and Motion Control Conference), Riga, Latvia, September 2004.
[13] H. Miranda, P. Cortés, J. I. Yuz, and J. Rodríguez, "Predictive torque control of induction machines based on state-space models," *IEEE Transactions on Industrial Electronics*, vol. 56, no. 6, pp. 1916–1924, June 2009.

12

Delay Compensation

12.1 Introduction

When control schemes based on model predictive control (MPC) are implemented experimentally, a large number of calculations are required, introducing a considerable time delay in the actuation. This delay can deteriorate the performance of the system if not considered in the design of the controller.

Compensation of the calculation delay has been considered in several works published to date [1–5]. Similar compensation methods have also been proposed for other predictive control schemes such as deadbeat control [6].

Another source of delay in these types of control schemes appears due to the need for future values of the reference variables in the cost function. Usually, the future reference is considered to be the same as the actual reference, which is a good assumption when the reference is a constant value or the sampling frequency is much higher than the frequency of the reference variable. However, during transients and with sinusoidal references, a delay between the controlled and reference variables appears. In order to eliminate this delay, the future reference variables need to be calculated. Some simple extrapolation methods for calculating the future reference variables are presented in this chapter.

12.2 Effect of Delay due to Calculation Time

The control of a three-phase inverter with a passive load (resistive–inductive) is used as an example application for explaining the effects of the delay due to calculation time and the delay compensation method. However, the same ideas are valid for all predictive control schemes.

The predictive current control scheme using MPC is shown in Figure 12.1 and consists of the following steps:

1. Measurement of the load currents.
2. Prediction of the load currents for the next sampling instant for all possible switching states.
3. Evaluation of the cost function for each prediction.

Predictive Control of Power Converters and Electrical Drives, First Edition. Jose Rodriguez and Patricio Cortes.

4. Selection of the switching state that minimizes the cost function.
5. Application of the new switching state.

The predictive control algorithm can also be represented as the flowchart presented in Figure 12.2. As can be seen in this figure, calculation of the predicted current and cost function is repeated as many times as there are available switching states, leading to a large number of calculations performed by the microprocessor.

In the case of current control, the cost function is defined as the error between the reference current and the predicted currents for a given switching state, and is expressed as

$$g = \left|i_\alpha^*(k+1) - i_\alpha^{\mathbf{p}}(k+1)\right| + \left|i_\beta^*(k+1) - i_\beta^{\mathbf{p}}(k+1)\right| \tag{12.1}$$

where i_α^* and i_β^* are the real and imaginary parts of the reference current vector, and $i_\alpha^{\mathbf{p}}$ and $i_\beta^{\mathbf{p}}$ are the real and imaginary parts of the predicted load current vector $\mathbf{i}^{\mathbf{p}}(k+1)$. The predicted load current vector is calculated using a discrete-time model of the load, which is a function of the measured currents $\mathbf{i}(k)$ and the inverter voltage (the actuation) $\mathbf{v}(k)$, and is expressed as

$$\mathbf{i}^{\mathbf{p}}(k+1) = \left(1 - \frac{RT_s}{L}\right)\mathbf{i}(k) + \frac{T_s}{L}\mathbf{v}(k) \tag{12.2}$$

where R and L are the load resistance and inductance, and T_s is the sampling time.

To graphically illustrate the predictive current control, only i_β is shown in Figure 12.3. This simplifies the example as the seven different voltage vectors produce only three different values for their β component and hence there are only three possible trajectories for i_β. In this figure, the dashed lines represent the predictions for i_β, as given by (12.2), and the solid line is the actual trajectory given by the application of the optimum voltages obtained by minimization of the cost function (12.1).

In the ideal case, the time needed for the calculations is negligible and the predictive control operates as shown in Figure 12.3. This ideal case is shown for comparison. The currents are measured at time t_k and the optimal switching state is calculated instantly. The switching state that minimizes the error at time t_{k+1} is selected and applied at time t_k. Then, the load current reaches the predicted value at t_{k+1}.

As the three-phase inverter has seven different voltage vectors, the predicted current (12.2) and the cost function (12.3) are calculated seven times. In this way, depending on the sampling frequency and the speed of the microprocessor used for the control, the

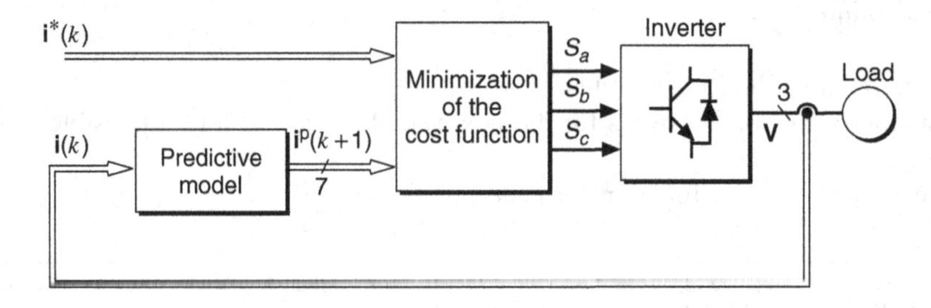

Figure 12.1 Predictive control scheme for a three-phase inverter

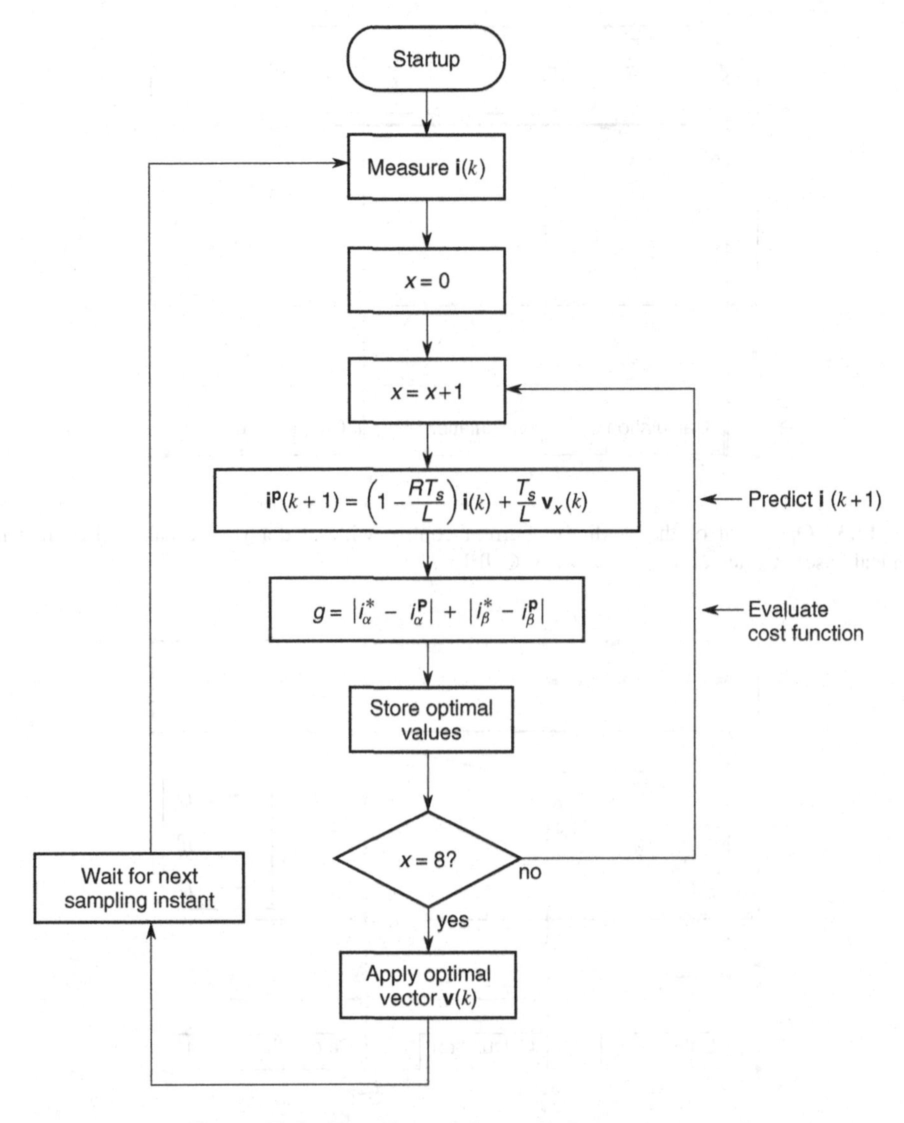

Figure 12.2 Flowchart of the predictive current control

time between measurement of the load currents and application of the new switching state can be considerable.

If the calculation time is significant compared to the sampling time, there will be a delay between the instant at which the currents are measured and the instant of application of the new switching state, as shown in Figure 12.4. During the interval between these two instants, the previous switching state will continue to be applied. As can be observed in the figure, the voltage vector selected using measurements at t_k will continue being applied after t_{k+1}, making the load current move away from the reference. The next actuation will be selected considering the measurements in t_{k+1} and will be applied near t_{k+2}. As a consequence of this delay, the load current will oscillate around its reference, increasing the current ripple.

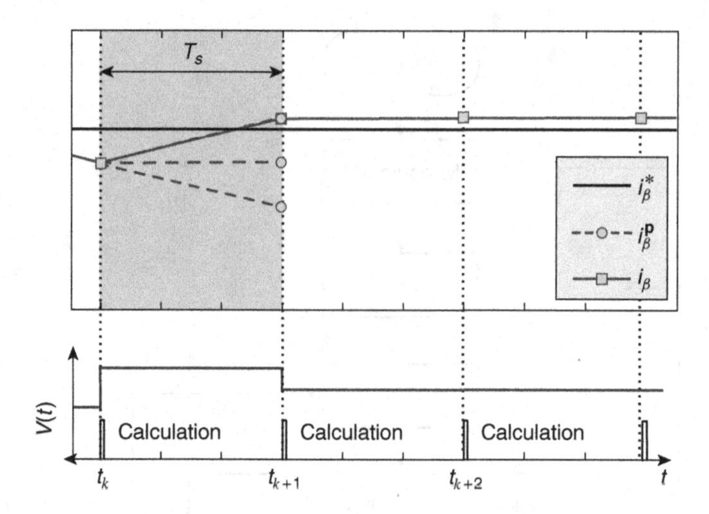

Figure 12.3 Operation of the predictive current control without delay. The calculation time is zero (ideal case) (Cortes *et al.*, forthcoming © IEEE)

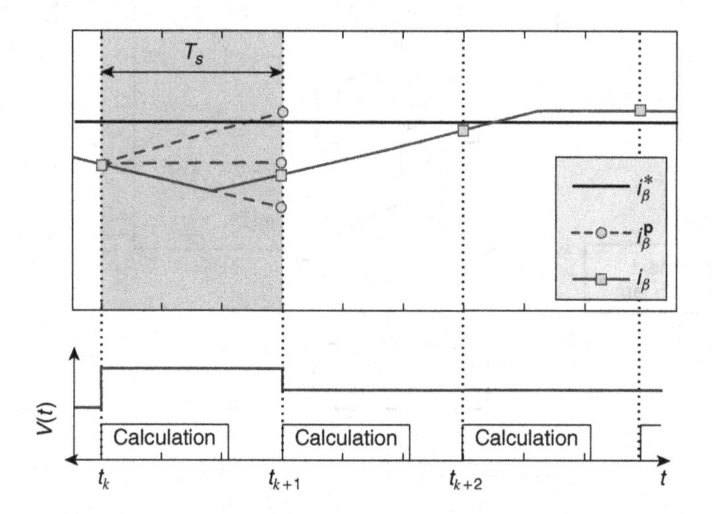

Figure 12.4 Operation of the predictive current control with delay and without compensation: long calculation time (real case) (Cortes *et al.*, forthcoming © IEEE)

12.3 Delay Compensation Method

A simple solution to compensate this delay is to take into account the calculation time and apply the selected switching state after the next sampling instant. In this way, the control algorithm is modified as follows:

1. Measurement of the load currents.
2. Application of the switching state (calculated in the previous interval).

3. Estimation of the value of the currents at time t_{k+1}, considering the applied switching state.
4. Prediction of the load currents for the next sampling instant t_{k+2} for all possible switching states.
5. Evaluation of the cost function for each prediction.
6. Selection of the switching state that minimizes the cost function.

The predictive control algorithm with delay compensation can also be represented as the flowchart presented in Figure 12.5. Compared to the original control algorithm shown in Figure 12.2, application of the new voltage vector is moved to the beginning, and the estimation of the currents at time t_{k+1} is added. Note that the estimation of the currents increases the calculation time, but only marginally because it must be calculated only once.

The operation of the predictive control with compensation delay is shown in Figure 12.6. Here, the measured currents and the applied switching state at time t_k are used in (12.2) to estimate the value of the load currents at time t_{k+1}. Then, this current is used as a starting point for the predictions for all switching states. These predictions are calculated using the load model shifted one step forward in time:

$$\mathbf{i}^{\mathrm{p}}(k+2) = \left(1 - \frac{RT_s}{L}\right)\hat{\mathbf{i}}(k+1) + \frac{T_s}{L}\mathbf{v}(k+1) \tag{12.3}$$

where $\hat{\mathbf{i}}(k+1)$ is the estimated current vector and $\mathbf{v}(k+1)$ is the actuation to be evaluated.

The cost function is modified for evaluation of the predicted currents $\mathbf{i}^{\mathrm{p}}(k+2)$, resulting in

$$g = |i_\alpha^*(k+2) - i_\alpha^{\mathrm{p}}(k+2)| + |i_\beta^*(k+2) - i_\beta^{\mathrm{p}}(k+2)| \tag{12.4}$$

and the switching state that minimizes this cost function is selected and stored to be applied at the next sampling instant.

Operation of the predictive current control method with a large delay due to the calculations is shown in Figure 12.7. It can be seen that the ripple in the load currents is considerable when the delay is not compensated. The delay compensation method reduces the ripple and operation is similar to the ideal case.

Note that cost functions (12.2) and (12.3) require future values of the reference currents $\mathbf{i}^*(k+1)$ and $\mathbf{i}^*(k+2)$, respectively. The calculation of these values is discussed in the next section.

12.4 Prediction of Future References

In the predictive control strategies presented throughout this book, the cost function is based on the future error, that is, the error between the predicted variable and the reference at the next sampling instant. This means that future references need to be known.

In general, future references are not known, so they need to be estimated. A very simple approach, based on the assumption that the sampling frequency is much higher than the frequency of the reference signal, considers that the future value of the reference is approximately equal to the present value of the reference.

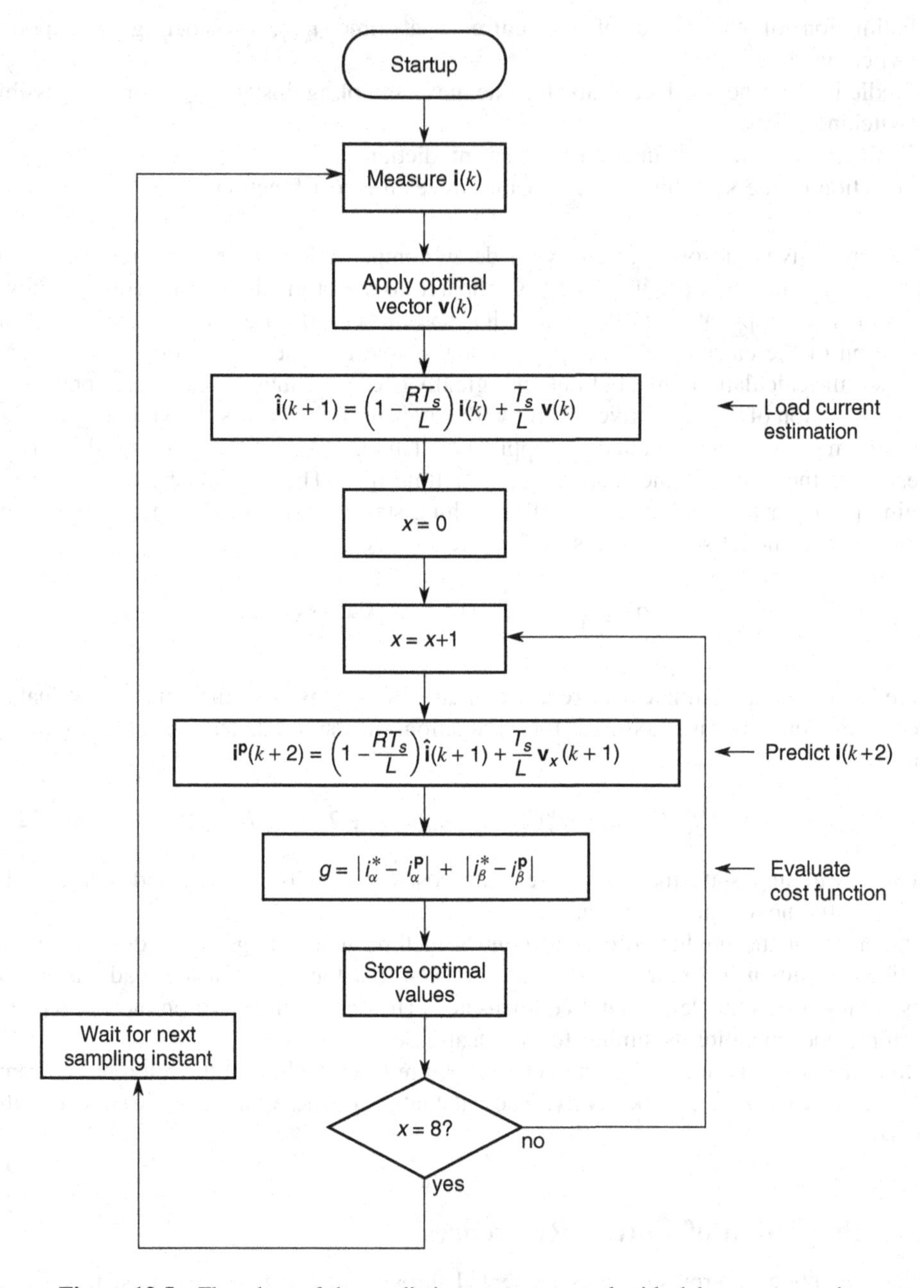

Figure 12.5 Flowchart of the predictive current control with delay compensation

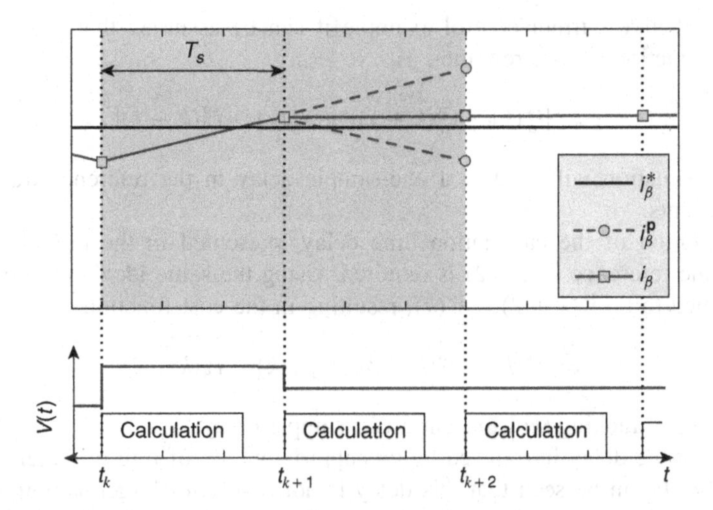

Figure 12.6 Operation of the predictive current control with delay and compensation: long calculation time (real case) (Cortes *et al.*, forthcoming © IEEE)

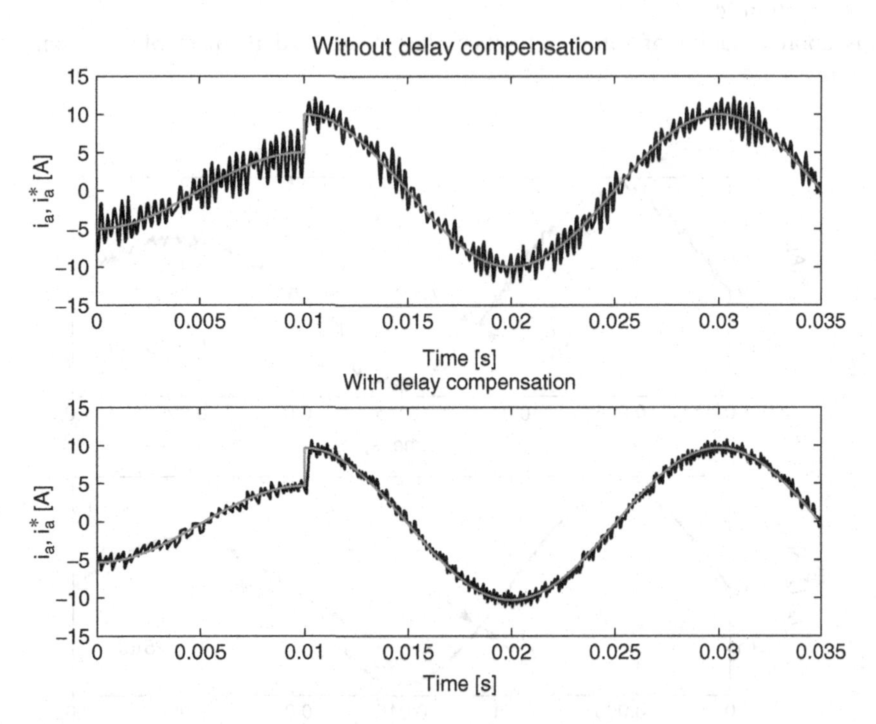

Figure 12.7 Predictive current control operation without and with delay compensation ($T_s = 50\,\mu s$) (Cortes *et al.*, forthcoming © IEEE)

For the predictive current control example it can be assumed that $\mathbf{i}^*(k+1) = \mathbf{i}^*(k)$, and the cost function can be rewritten as

$$g = |i^*_\alpha(k) - i^p_\alpha(k+1)| + |i^*_\beta(k) - i^p_\beta(k+1)| \tag{12.5}$$

This approximation will lead to a one-sample delay in the reference tracking of the reference currents.

If compensation of the calculation time delay, presented in the previous section, is considered, the reference $\mathbf{i}^*(k+2)$ is required. Using the same idea, the future reference can be assumed to be $\mathbf{i}^*(k+2) = \mathbf{i}^*(k)$, resulting in the cost function

$$g = |i^*_\alpha(k) - i^p_\alpha(k+2)| + |i^*_\beta(k) - i^p_\beta(k+2)| \tag{12.6}$$

and the reference tracking will present a two-sample delay.

The effect of the delay introduced by this approximation of future references is shown in Figure 12.8. It can be seen that this delay is noticeable for larger sampling times like $T_s = 100\,\mu\text{s}$, but it is not visible when the sampling time is smaller.

It is common to use smaller sampling times in predictive control schemes, so this approach is reasonable is those cases. When the references are constant at steady state operation, this approach has no negative effects, and the two-sample delay can be observed only during transients.

This approximation of future references has been used in most of the examples in this book.

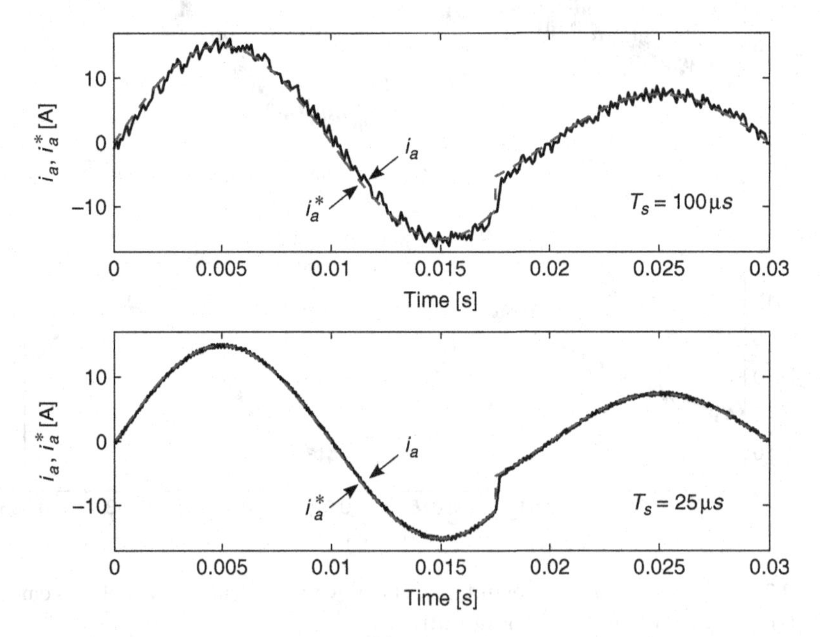

Figure 12.8 Load current and load current reference for predictive current control using $\mathbf{i}^*(k+2) = \mathbf{i}^*(k)$ for sampling times $T_s = 100\,\mu\text{s}$ and $T_s = 25\,\mu\text{s}$

12.4.1 Calculation of Future References Using Extrapolation

For sinusoidal references and large sampling times, the use of extrapolation methods for the reference can compensate the delay in the reference tracking in predictive control schemes.

A possible solution is to calculate the one-step-ahead prediction using the actual current reference in the nth-order formula of the Lagrange extrapolation [7] by

$$\hat{\mathbf{i}}^*(k+1) = \sum_{l=0}^{n} (-1)^{n-l} \begin{bmatrix} n+1 \\ l \end{bmatrix} \mathbf{i}^*(k+l-n) \tag{12.7}$$

For sinusoidal references, $n = 2$ or higher is recommended [7].

Using this extrapolation formula, the future reference $\mathbf{i}^*(k+1)$ can be predicted, for $n = 2$, with

$$\hat{\mathbf{i}}^*(k+1) = 3\mathbf{i}^*(k) - 3\mathbf{i}^*(k-1) + \mathbf{i}^*(k-2) \tag{12.8}$$

Calculation of the future reference $\mathbf{i}^*(k+2)$ is required when cost function (12.4) is considered. This estimate can be calculated by shifting forward (12.8), giving

$$\hat{\mathbf{i}}^*(k+2) = 3\hat{\mathbf{i}}^*(k+1) - 3\mathbf{i}^*(k) + \mathbf{i}^*(k-1) \tag{12.9}$$

and, by substituting (12.8) into (12.9), the future reference can be calculated using only present and past values of the reference current. The resulting expression for the calculation of $\mathbf{i}^*(k+2)$ is

$$\hat{\mathbf{i}}^*(k+2) = 6\mathbf{i}^*(k) - 8\mathbf{i}^*(k-1) + 3\mathbf{i}^*(k-2) \tag{12.10}$$

The reference current $\mathbf{i}^*(k)$ and the estimated future reference $\hat{\mathbf{i}}^*(k+1)$, calculated using (12.8), are shown in Figure 12.9. It can be seen that a good estimate of the future value of the current is obtained at steady state operation. However, some peaks appear during step changes of the reference.

The estimated future reference $\hat{\mathbf{i}}^*(k+2)$ is also shown in Figure 12.9 and presents a similar behavior than $\hat{\mathbf{i}}^*(k+1)$, but the peak during the step change is higher.

12.4.2 Calculation of Future References Using Vector Angle Compensation

Taking into account the vectorial representation of the variables of a three-phase system, it is possible to implement an estimate of the future reference by considering the change in the vector angle during one sample time.

For example, the load current reference vector $\mathbf{i}^*$ can be described by its magnitude I^* and angle θ:

$$\mathbf{i}^*(k) = I^*(k)e^{j\theta(k)} \tag{12.11}$$

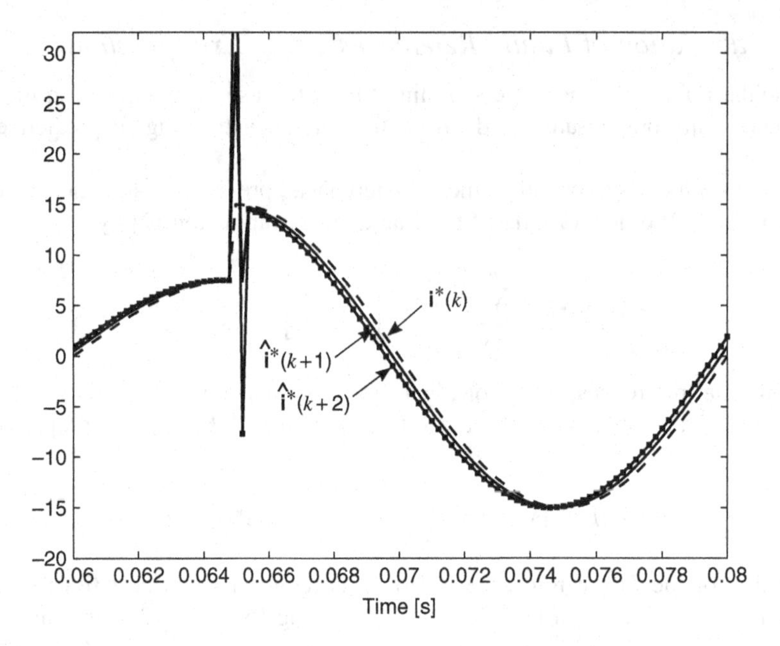

Figure 12.9 Future reference using extrapolation for a sinusoidal reference with a step change in amplitude

At steady state, it can be assumed that this vector rotates at an angular speed ω and that the magnitude remains constant. In this way, the angle of the reference vector for time t_{k+1} can be estimated as

$$\theta(k+1) = \theta(k) + \omega T_s \tag{12.12}$$

where T_s is the sampling time.

Considering (12.12) and $I^*(k+1) = I^*(k)$, the value of the future reference vector can be estimated as

$$\mathbf{i}^*(k+1) = I^*(k+1)e^{j\theta(k+1)} = I^*(k)e^{j(\theta(k)+\omega T_s)} \tag{12.13}$$

and inserting (12.11) into (12.13) results in

$$\mathbf{i}^*(k+1) = \mathbf{i}^*(k)e^{j\omega T_s} \tag{12.14}$$

The same procedure can be used for the estimation of $\mathbf{i}^*(k+2)$, assuming

$$\theta(k+2) = \theta(k) + 2\omega T_s \tag{12.15}$$

and $I^*(k+2) = I^*(k)$, resulting in

$$\mathbf{i}^*(k+2) = \mathbf{i}^*(k)e^{2j\omega T_s} \tag{12.16}$$

A vector diagram of the reference vector and future references calculated using this method is shown in Figure 12.10.

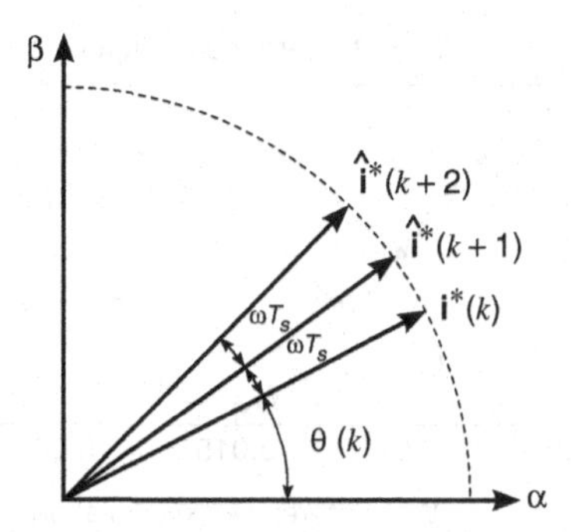

Figure 12.10 Current reference vector and future references calculated using vector angle compensation

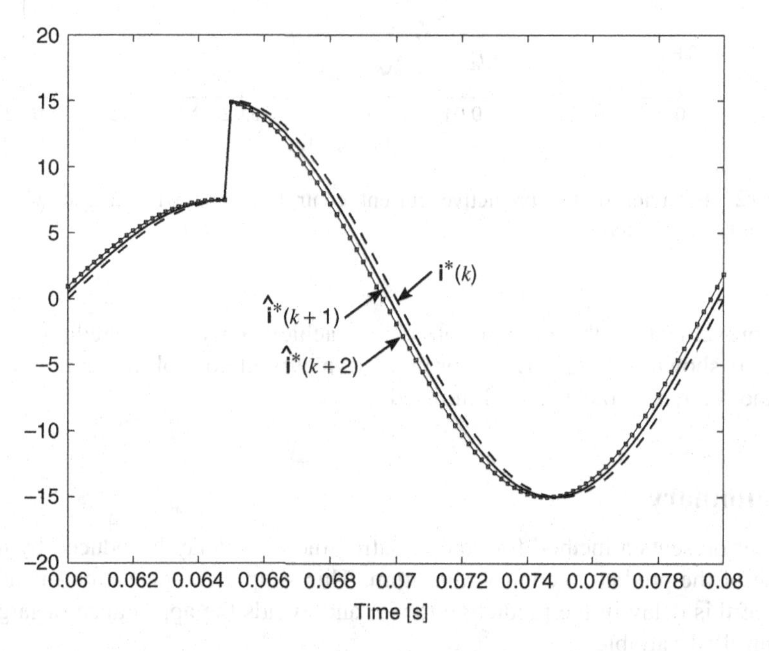

Figure 12.11 Future references using vector angle compensation for a sinusoidal reference with a step change in amplitude

The reference current $\mathbf{i}^*(k)$ and the estimated future references $\hat{\mathbf{i}}^*(k+1)$ and $\hat{\mathbf{i}}^*(k+2)$, calculated using (12.14) and (12.16), respectively, are shown in Figure 12.11. It can be observed that a good estimate of the future value of the current is obtained at steady state operation. The behavior of the estimated references during step changes of the reference is better than that presented in Figure 12.9 for the extrapolation method.

Estimation of future references using vector angle compensation cannot be used for single-phase systems.

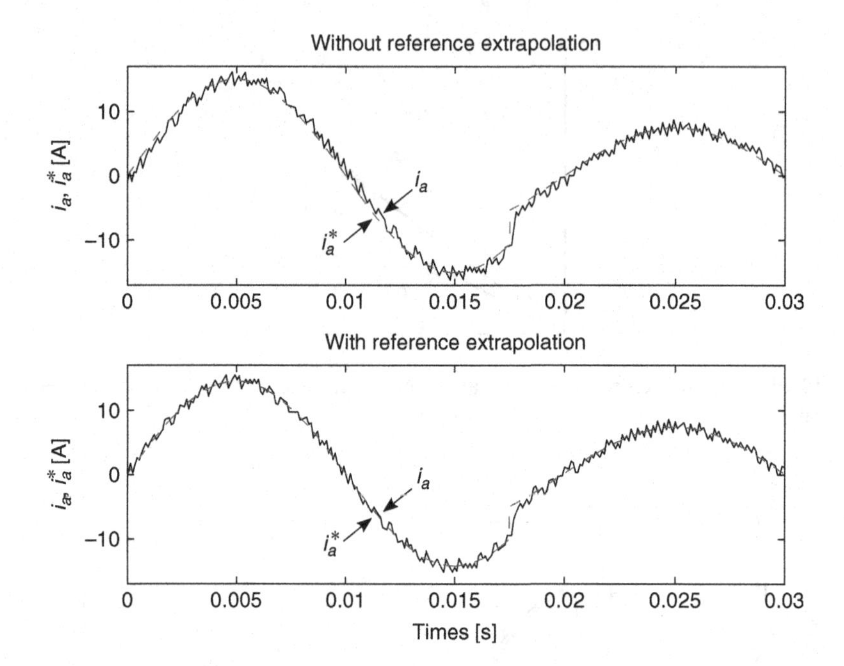

Figure 12.12 Behavior of the predictive current control method without and with reference extrapolation ($T_s = 100\,\mu s$)

The improvement in the reference tracking, achieved with the prediction of future references, is shown in Figure 12.12 for predictive current control. It can be seen that the delay in the reference tracking is eliminated.

12.5 Summary

This chapter presents a method for compensating the time delay introduced by the calculation time of the predictive control algorithm. This simple compensation method allows inclusion of this delay in the predictive model and avoids the appearance of large ripples in the controlled variable.

Considering that in some cases the future value of the references is required, two extrapolation methods for estimating future references are also presented in this chapter.

References

[1] P. Cortés, G. Ortiz, J. I. Yuz *et al.*, "Model predictive control of an inverter with output LC filter for UPS applications," *IEEE Transactions on Industrial Electronics*, vol. 56, no. 6, pp. 1875–1883, June 2009.

[2] P. Cortés, J. Rodríguez, D. E. Quevedo, and C. Silva, "Predictive current control strategy with imposed load current spectrum," *IEEE Transactions on Power Electronics*, vol. 23, no. 2, pp. 612–618, March 2008.

[3] P. Cortés, J. Rodríguez, P. Antoniewicz, and M. Kazmierkowski, "Direct power control of an AFE using predictive control," *IEEE Transactions on Power Electronics*, vol. 23, no. 5, pp. 2516–2523, September 2008.

[4] H. Miranda, P. Cortés, J. I. Yuz, and J. Rodríguez, "Predictive torque control of induction machines based on state-space models," *IEEE Transactions on Industrial Electronics*, vol. 56, no. 6, pp. 1916–1924, June 2009.

[5] M. Arahal, F. Barrero, S. Toral, M. Duran, and R. Gregor, "Multi-phase current control using finite-state model-predictive control," *Control Engineering Practice*, vol. 17, no. 5, pp. 579–587, 2009.

[6] H. Abu-Rub, J. Guzinski, Z. Krzeminski, and H. Toliyat, "Predictive current control of voltage-source inverters," *IEEE Transactions on Industrial Electronics*, vol. 51, no. 3, pp. 585–593, June 2004.

[7] O. Kukrer, "Discrete-time current control of voltage-fed three-phase PWM inverters," *IEEE Transactions on Industrial Electronics*, vol. 11, no. 2, pp. 260–269, March 1996.

13

Effect of Model Parameter Errors

13.1 Introduction

An important characteristic of model predictive control (MPC) is the explicit use of the system models for selecting the optimal actuations. Considering that the parameter values may vary in some systems while in other cases it is difficult to get a precise value of the parameters, it is important to evaluate how MPC schemes behave in the presence of errors in the model parameters.

Due to the nonlinear nature of the predictive control scheme presented in this book, it is not possible to perform a simple analytical study in order to evaluate the behavior of predictive control in the presence of model parameter errors. This chapter presents a simple empirical approach to test the effect of model parameter errors at steady state and transient operation of the system. As an example, the current control of a three-phase inverter is considered. As references for a comparison of the results, a classical control scheme based on PI controllers in rotating coordinates with PWM and the well-known deadbeat controller, has been selected.

13.2 Three-Phase Inverter

The three-phase inverter shown in Figure 13.1 will be considered for the comparisons presented in this chapter.

A resistive–inductive load is considered, where the following simple model describes the dynamic behavior of the load current:

$$\mathbf{v} = R\mathbf{i} + L\frac{d\mathbf{i}}{dt} \tag{13.1}$$

This simple model will be used for the three control methods explained in the subsequent sections.

Predictive Control of Power Converters and Electrical Drives, First Edition. Jose Rodriguez and Patricio Cortes.

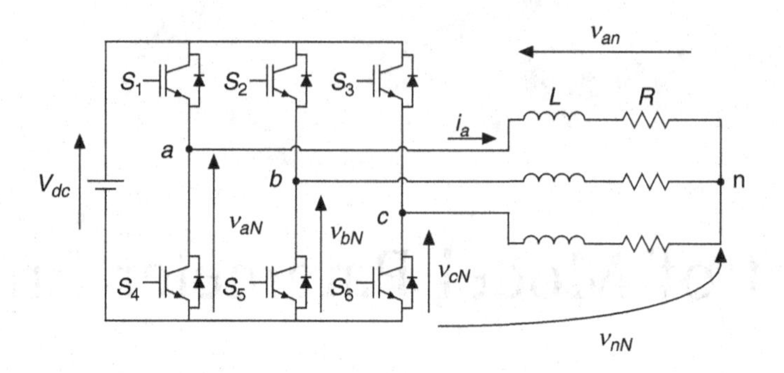

Figure 13.1 Three-phase inverter with resistive–inductive load

13.3 Proportional–Integral Controllers with Pulse Width Modulation

13.3.1 Control Scheme

The use of PI controllers with a pulse width modulator is the method with the greatest development and the most established in the literature [1–3]. Here, the current error is compensated by the PI controllers, which generate the reference voltages that are inputs to the modulator.

This control method can be implemented by considering a stationary or synchronous reference frame for the load current vector [4]. In a stationary reference frame, the controlled currents are sinusoidal signals and the PI controllers are not able to reduce the error to zero. In a synchronous rotating d–q reference frame the load current components are DC quantities and the PI controllers reduce the error to zero at steady state, hence in this work a synchronous frame regulator is used. Considering the synchronous d–q coordinates [5], the load model (13.1) can be expressed as follows:

$$v_d = Ri_d + L\frac{di_d}{dt} - j\omega Li_q \tag{13.2a}$$

$$v_q = Ri_q + L\frac{di_q}{dt} + j\omega Li_d \tag{13.2b}$$

where ω is the angular frequency of the rotating frame, v_d and v_q are the components of the voltage vector generated by the inverter in the rotating frame and i_d and i_q are the components of the load current vector in the rotating frame.

The equations (13.2a) and (13.2b) are the mathematical model of the system, which is used in the design of the PI regulators. The selection of PI regulator parameters is a key factor and must take into account the requirements of the system under control. There are some tuning techniques that have been traditionally used to select the PI regulator parameters. Hence a traditional tuning technique that has been adapted for microprocessor control is used [6].

The control scheme for current control is shown in Figure 13.2. The measured currents are transformed to the rotating coordinates frame and then the error is calculated by considering the corresponding reference signals. The error signal is the input to the PI controller that generates the reference voltages in rotating coordinates. This reference

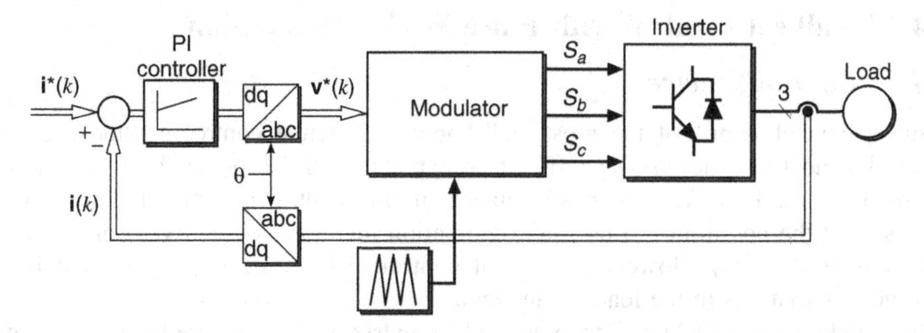

Figure 13.2 PI control with PWM block diagram

voltage is transformed to the stationary reference frame and enters the modulator that generates the required gate signals for the inverter.

13.3.2 Effect of Model Parameter Errors

In order to analyze the behavior of the system under variation of its parameters, simulations and experimental tests will be performed by varying only one of the parameters and keeping the other one at its nominal value. The range is 3 to 15 mH in the case of the inductance value and 5 to 20 Ω in the case of the resistance value. The nominal values are 7 mH and 10 Ω respectively.

The effect of variation of the load parameter values has been investigated in terms of the closed loop stability. The method used in previous works is to form the closed loop characteristic equation and then plot the root locus [7]. The same method will be adopted in this work for the case of PI control with PWM and deadbeat control.

Figure 13.3 shows the root locus plot when the parameter values have been varied within the previously defined range. The PI regulator has been designed according to the guidelines established in [6]. As can be observed, the system remains stable within the range where both parameters are varied.

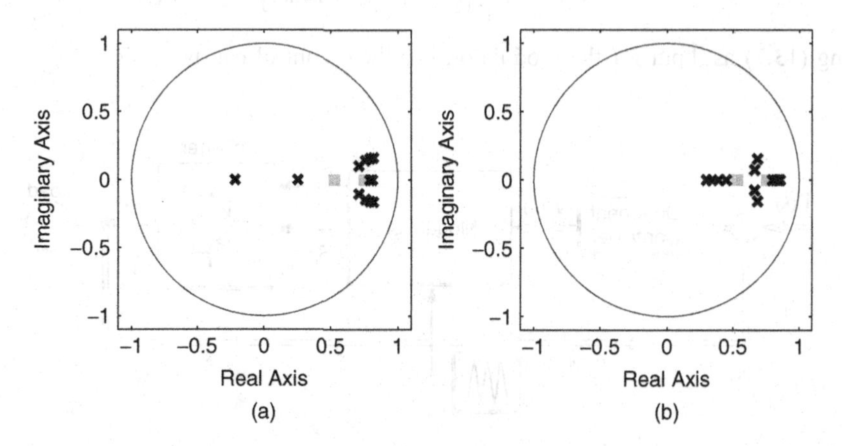

Figure 13.3 Root locus using PI control. (a) Inductance value variation. (b) Resistance value variation

13.4 Deadbeat Control with Pulse Width Modulation

13.4.1 Control Scheme

Deadbeat control is one of the most well-known predictive control methods, and was proposed almost two decades ago [8, 9]. A typical control scheme for this method is shown in Figure 13.4. The main advantages of this control scheme are fast dynamic response and the possibility to use any modulation method such as PWM or space vector modulation (SVM). However, deadbeat control in its basic implementation is very sensitive to variations in the load parameters.

The simple model (13.1) can be expressed in matrix notation where the currents in the stationary frame are considered as state variables as follows:

$$\begin{bmatrix} \dot{i}_\alpha \\ \dot{i}_\beta \end{bmatrix} = \begin{bmatrix} -R/L & 0 \\ 0 & -R/L \end{bmatrix} \begin{bmatrix} i_\alpha \\ i_\beta \end{bmatrix} + \begin{bmatrix} 1/L & 0 \\ 0 & 1/L \end{bmatrix} \begin{bmatrix} v_\alpha \\ v_\beta \end{bmatrix} \tag{13.3}$$

where i_α and i_β are the components of the load current vector $\mathbf{i}$, and v_α and v_β are the components of the inverter voltage vector $\mathbf{v}$.

Under the assumption that the variables are constant between sampling instants, the system (13.3) can be discretized as follows:

$$\mathbf{i}(k+1) = \Phi\mathbf{i}(k) + \Gamma\mathbf{v}(k) \tag{13.4}$$

where

$$\Phi = e^{-(R/L)T} \tag{13.5}$$

$$\Gamma = \int_0^T e^{-(R/L)\tau} d\tau \cdot \frac{1}{L} \tag{13.6}$$

and T is the sampling time.

Based on the discrete-time model (13.4), the reference voltage vector can be obtained in order to achieve zero current error at the next sampling instant:

$$\mathbf{v}^*(k) = \frac{1}{\Gamma}\left[\mathbf{i}^*(k+1) - \Phi\mathbf{i}(k)\right] \tag{13.7}$$

Using (13.7) as input for the modulator, deadbeat control can be achieved.

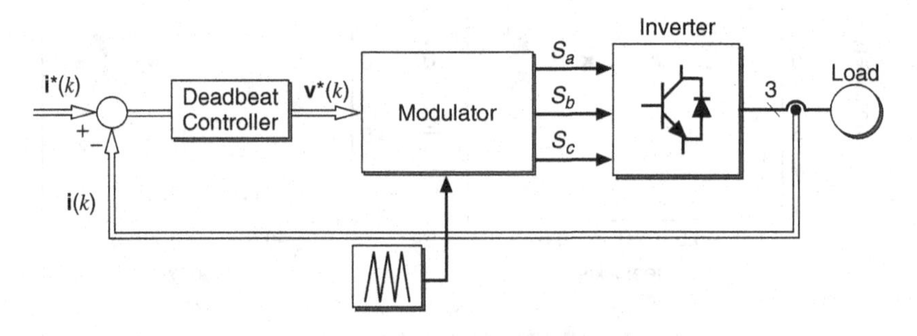

Figure 13.4 Deadbeat control with PWM block diagram

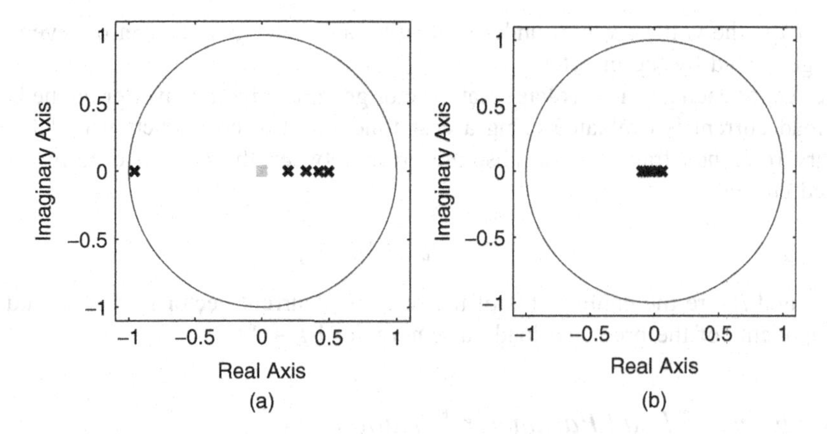

Figure 13.5 Root locus using deadbeat control. (a) Inductance value variation. (b) Resistance value variation

13.4.2 Effect of Model Parameter Errors

The effect of variations of the parameter values within the predefined range is shown in Figure 13.5 in the form of a root locus plot. It can be observed that the system remains stable for the whole range. However, it can also be observed that there is a particular case where one of the system closed loop poles is near the instability zone. This situation occurs when the value of the inductance reaches its minimum value.

13.5 Model Predictive Control

MPC is a type of predictive control where a model of the system is used in order to predict the behavior of the variables under control. The optimal input sequence is selected by minimizing a cost function, which defines the desired behavior of the system. A simple and effective implementation of MPC for current control in a voltage source inverter is presented in [10]. The block diagram for this method is shown in Figure 13.6. A detailed explanation of this control method can be found in Chapter 4.

The predictive model used for calculating the predicted currents is obtained by discretization (using the Euler method) of (13.3) for a sampling time T, and is expressed as

$$\mathbf{i}(k+1) = \left(1 - \frac{RT}{L}\right)\mathbf{i}(k) + \frac{T}{L}\mathbf{v}(k) \tag{13.8}$$

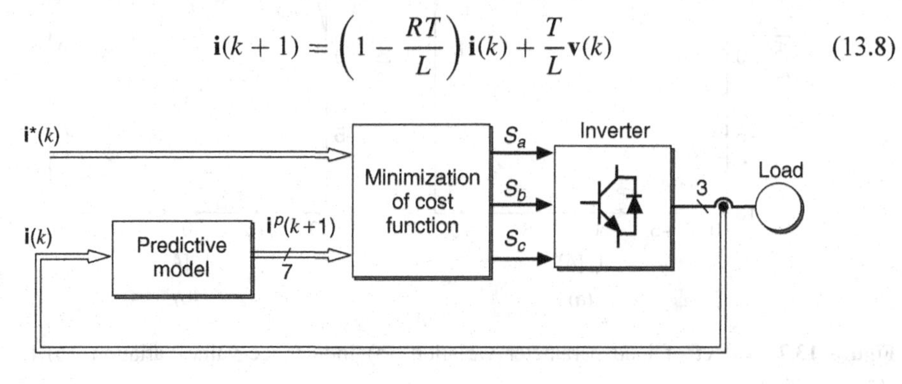

Figure 13.6 MPC block diagram

where $\mathbf{v}(k)$ is the voltage vector under evaluation and belongs to the set of seven voltage vectors generated by the inverter.

The effect of each possible voltage state vector generated by the inverter on the behavior of the load current is evaluated using a cost function. The cost function is defined in a stationary reference frame as the absolute error between the reference current and the predicted current:

$$g = |i_\alpha^* - i_\alpha^p| + |i_\beta^* - i_\beta^p| \tag{13.9}$$

where i_α^* and i_β^* are the components of the reference current vector $\mathbf{i}^*$, and i_α^p and i_β^p are the components of the predicted load current vector $\mathbf{i}(k+1)$.

13.5.1 Effect of Load Parameter Variation

Due to the nonlinear nature of this implementation of MPC, a root locus analysis is not possible and a closed loop stability analysis is more complex. For this reason, an analysis of the variation of the parameters will be performed in terms of the performance of the controller, rather than the effect on the stability of the closed loop.

The predictive model (13.8) can be rewritten as follows:

$$\mathbf{i}(k+1) = K_i\mathbf{i}(k) + K_v\mathbf{v}(k) \tag{13.10}$$

where $K_i = (1 - RT/L)$ and $K_v = T/L$.

Note in (13.10) that the predicted current has two components, the first one in the same direction as the actual current vector, and the another one in the direction of the seven possible voltage vectors. A variation in the resistance value only affects the weighting factor K_i; on the other hand, a variation in the inductance value affects both factors K_v and K_i.

It is possible to plot the effect of load parameter variations in the predicted currents at steady state. It can be seen in Figure 13.7 that variations in the load inductance produce

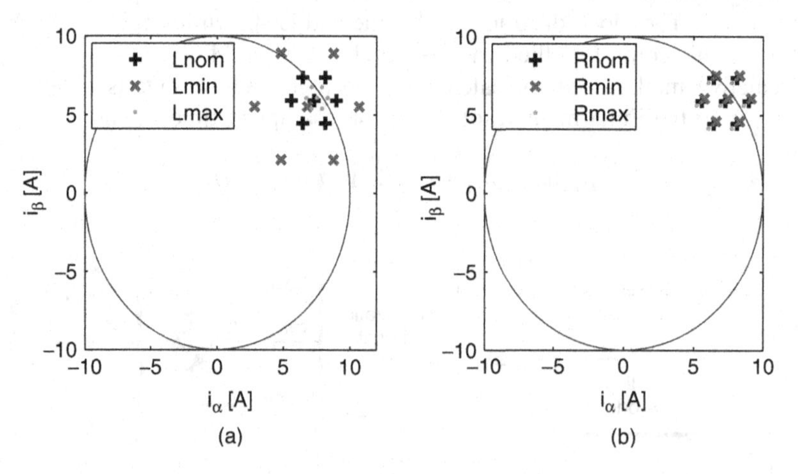

Figure 13.7 Effect of load parameter variation. (a) Inductance value variation. (b) Resistance value variation

variations in the predicted load current vector change $\Delta\mathbf{i}$. This will produce a variation in the current ripple and a small variation in the current magnitude, because both weighting factors are affected by a change in the value of the inductance, but the effect is more noticeable in K_v. On the other hand, variations in the load resistance produce a radial displacement of the predicted currents, resulting in variations in the predicted load current amplitude, because only K_i is affected when the value of the resistance is varied.

13.6 Comparative Results

Simulations were performed for the three methods using MATLAB/Simulink, with nominal system parameters $R = 10\,\Omega$, $L = 7\,\text{mH}$, and $V_{dc} = 540\,\text{V}$. For the PI controller and

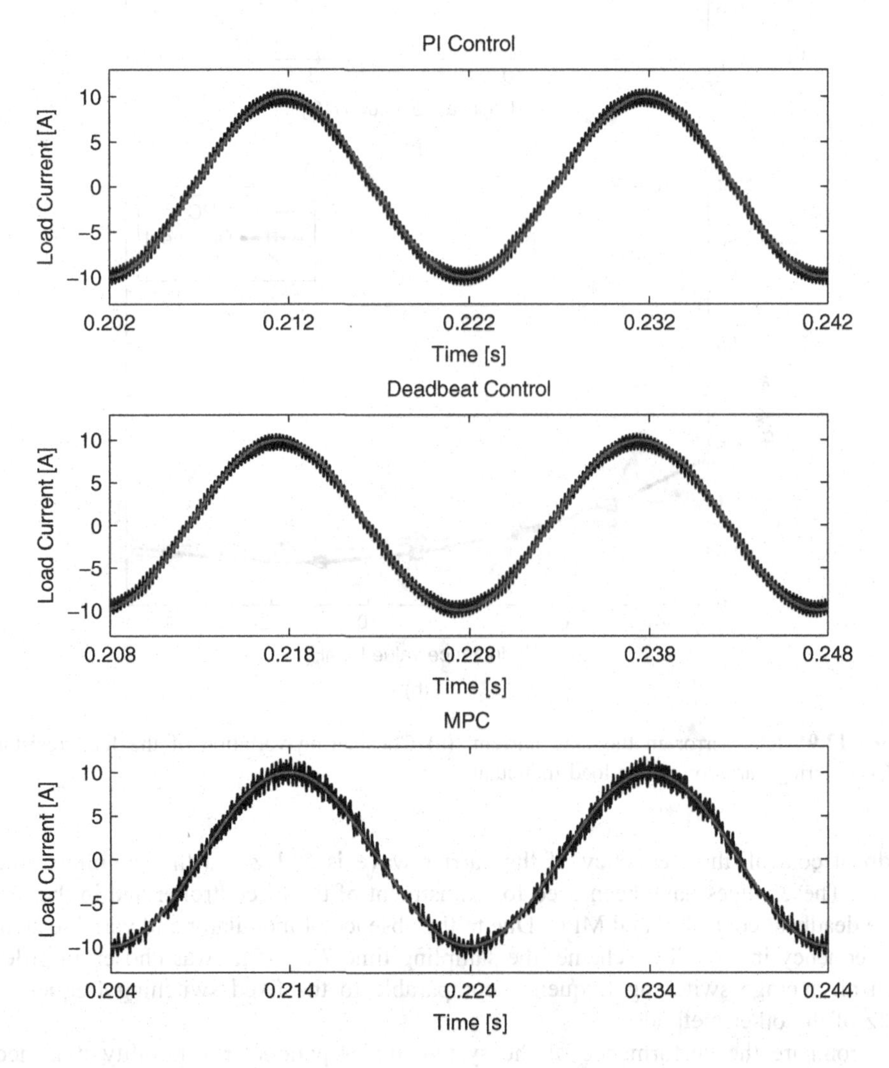

Figure 13.8 Steady state waveforms of the load current, with nominal values of the parameters

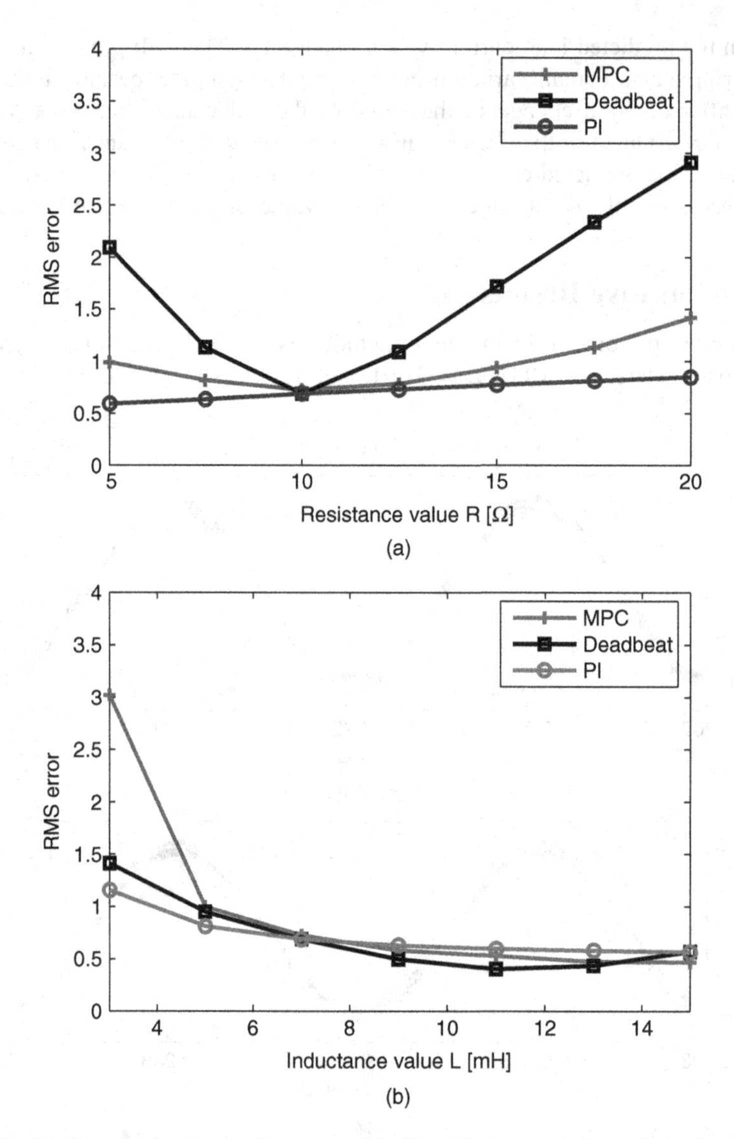

Figure 13.9 RMS error in the load current. (a) Considering variation of the load resistance. (b) Considering variation of the load inductance

deadbeat control, the frequency of the carrier wave is 5 kHz and the sampling time is 200 μs. These values have been used for adjustment of the PI controller and in the model of the deadbeat controller and MPC. Due to the absence of modulator and variable switching frequency in the MPC scheme, the sampling time $T = 45\,\mu s$ was chosen in order to reach an average switching frequency comparable to the fixed switching frequency of 5 kHz of the other methods.

To compare the performance of the system under parameter variability it is necessary to choose a performance index to evaluate significant aspects of the system under

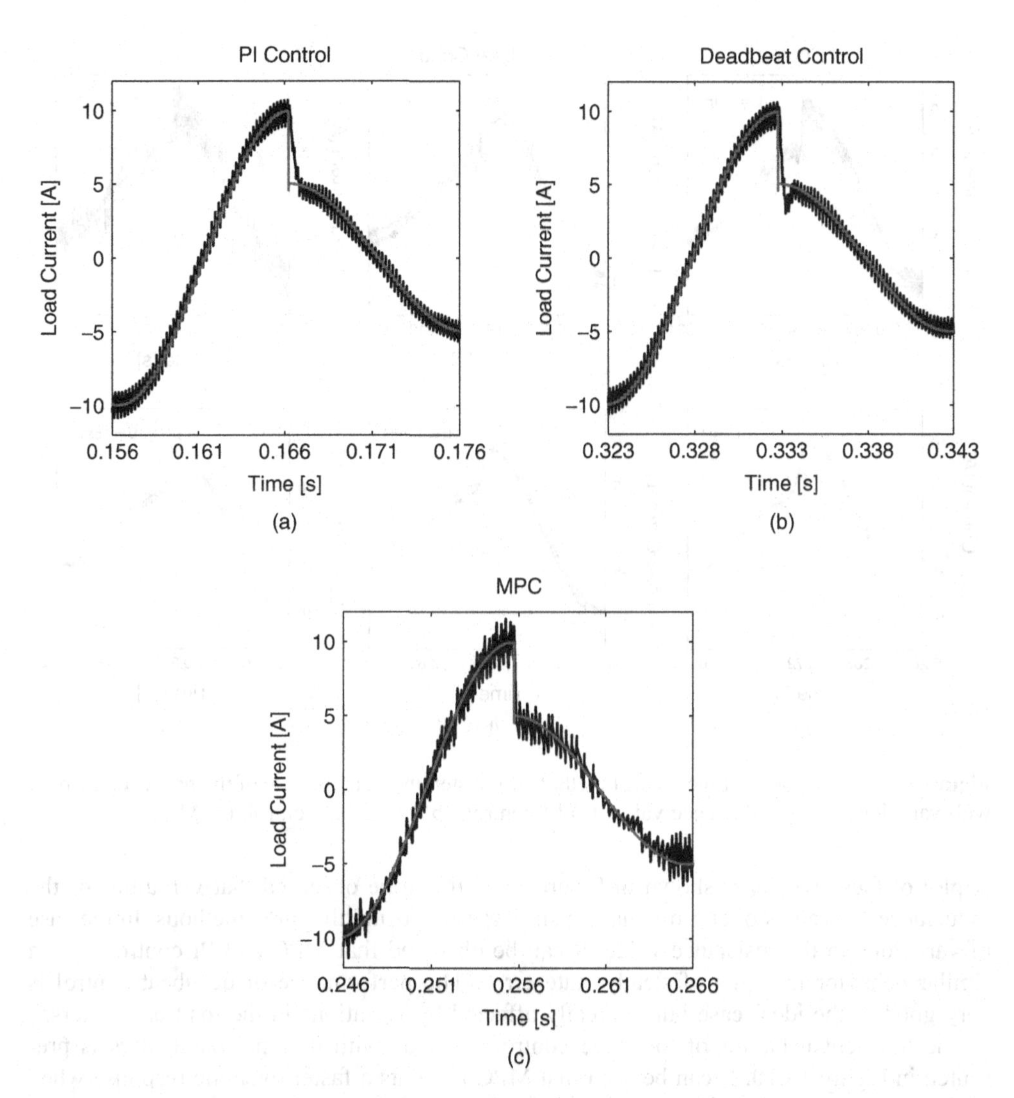

Figure 13.10 Response of the control methods for a step in the amplitude of the reference current, with nominal values of the load parameters. (a) PI Control. (b) Deadbeat Control. (c) MPC

control [11]. According to [12], the RMS current error was chosen because this index is suitable for evaluating how exactly the real current follows the reference current instantaneously.

Simulations were carried out to investigate the steady state performance of the system under the three control methods. The current waveforms at steady state with nominal parameters are shown in Figure 13.8. The three methods show good results, even in the case of MPC where a small increment in the current ripple is noticeable, compared to the other two methods.

The performance of each control method is evaluated by calculating the RMS error of the load currents for variations of the load inductance value and the load resistance value.

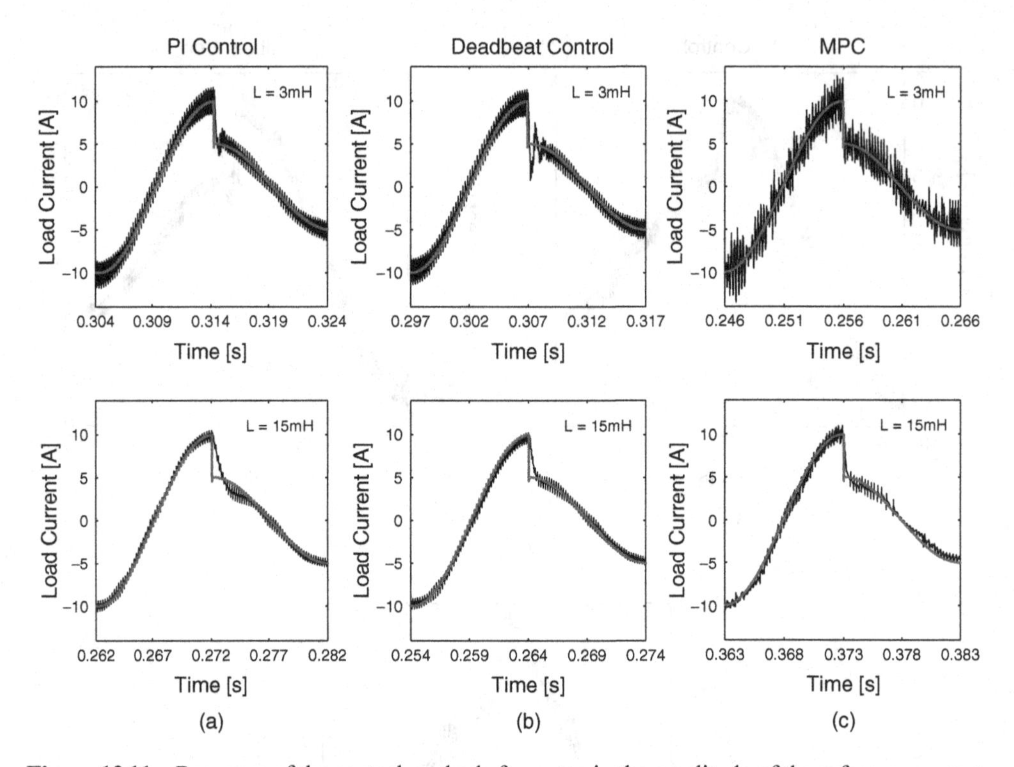

Figure 13.11 Response of the control methods for a step in the amplitude of the reference current, with variations in the inductance value. (a) PI control. (b) Deadbeat control. (c) MPC

A plot of these results is shown in Figure 13.9. It can be observed that variations of the inductance for values over 5 mH have a similar effect using the three methods. In the case of variations in the resistance value, it can be observed that MPC and PI control have a similar behavior in terms of steady state error. The performance of deadbeat control is very good in the ideal case but is greatly affected by variations in the load parameters.

The transient behavior of the three control methods with nominal parameters is presented in Figure 13.10. It can be seen that MPC presents a faster dynamic response when a step in the amplitude of the reference current is applied, compared to the other two control methods.

In order to evaluate the dynamic response of the system under variations of the parameters, the same test as explained before is replicated. Figures 13.11 and 13.12 show the effects when inductance and resistance values are changed respectively. It can be seen that the dynamic performance of the PI controller is slightly affected by load parameter variations, presenting a slower response in some cases. Deadbeat control reveals oscillations with inductance changes and large amplitude errors when the load resistance is different than the one used in the model. As was pointed out before in the root locus plot (Figure 13.5), with a small value for the inductance, the root of the system is very close to the zone of instability, generating oscillations when a step change occurs. MPC shows changes in the ripple for inductance variations and small amplitude errors with resistance variations. However, the fast dynamic response is almost unchanged.

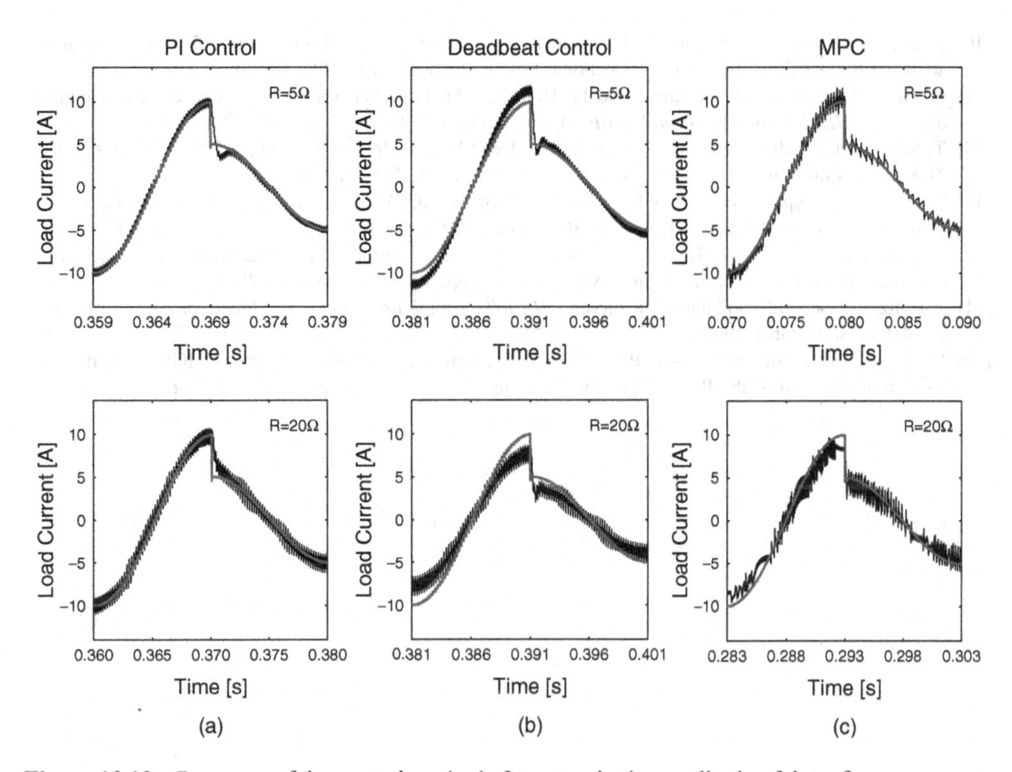

Figure 13.12 Response of the control methods for a step in the amplitude of the reference current, with variations in the resistance value. (a) PI control. (b) Deadbeat control. (c) MPC

13.7 Summary

The effect of model parameter errors on the performance of MPC is presented in this chapter. In order to avoid complex theoretical analysis, the performance of MPC is evaluated in comparison to two other well-known control methods.

It is shown that variations in the system parameter values can affect the performance of MPC in terms of RMS error of the load currents; however, its dynamic performance is not affected.

References

[1] M. P. Kaźmierkowski and H. Tunia, *Automatic Control of Inverter-Fed Drives*. Elsevier, 1994.

[2] M. P. Kaźmierkowski, R. Krishnan, and F. Blaabjerg, *Control in Power Electronics*. Academic Press, 2002.

[3] B. K. Bose, *Modern Power Electronics and AC Drives*. Prentice Hall, 2002.

[4] M. Kazmierkowski and L. Malesani, "Current control techniques for three-phase voltage-source PWM converters: a survey," *IEEE Transactions on Industrial Electronics*, vol. 45, no. 5, pp. 691–703, October 1998.

[5] T. M. Rowan and R. J. Kerkman, "A new synchronous current regulator and an analysis of current-regulated PWM inverters," *IEEE Transactions on Industry Applications*, vol. IA-22, no. 4, pp. 678–690, July 1986.

[6] J. Suul, M. Molinas, L. Norum, and T. Undeland, "Tuning of control loops for grid connected voltage source converters," in IEEE 2nd International PECon 2008, pp. 797–802, December 2008,

[7] G. Bode, P. C. Loh, M. Newman, and D. Holmes, "An improved robust predictive current regulation algorithm," *IEEE Transactions on Industry Applications*, vol. 41, no. 6, pp. 1720–1733, November 2005.

[8] T. Kawabata, T. Miyashita, and Y. Yamamoto, "Dead beat control of three phase PWM inverter," *IEEE Transactions on Power Electronics*, vol. 5, no. 1, pp. 21–28, January 1990.

[9] T. Habetler, "A space vector-based rectifier regulator for AC/DC/AC converters," *IEEE Transactions on Power Electronics*, vol. 8, no. 1, pp. 30–36, January 1993.

[10] J. Rodríguez, J. Pontt, C. Silva *et al.*, "Predictive current control of a voltage source inverter," *IEEE Transactions on Industrial Electronics*, vol. 54, no. 1, pp. 495–503, February 2007.

[11] J. Holtz, "Pulsewidth modulation–a survey," *IEEE Transactions on Industrial Electronics*, vol. 39, no. 5, pp. 410–420, October 1992.

[12] D.-C. Lee, S.-K. Sul, and M.-H. Park, "Comparison of AC current regulators for IGBT inverter," in Conference Record of the Power Conversion Conference. Yokohama, pp. 206–212, April 1993.

Appendix A

Predictive Control Simulation – Three-Phase Inverter

In this appendix three different power converter topologies, discussed previously in the book, are analyzed in depth from a simulation implementation perspective. The objective of the appendix is to give the reader the necessary tools to understand and replicate the implementation of predictive control algorithms using a simulation environment (MATLAB®/Simulink® in this particular case). Simulation is a key stage in predictive control design, since it provides valuable information on the control system performance which is needed to adjust control parameters such as weighting factors in the cost function. In addition, simulation is a preliminary validation required prior to experimenting on a real prototype. The case studies analyzed in this appendix are selected to range from simple to more advanced design considerations in predictive control. They also cover most of the elements and tools required to carry out all the controllers presented in this book.

This appendix does not review concepts presented previously in the book, and focuses mainly on implementing the simulation. Therefore, reading the theoretical and conceptual chapters related to the topologies addressed in this appendix is recommended.

A.1 Predictive Current Control of a Three-Phase Inverter

One of the most common converter topologies found in industry is the three-phase voltage source inverter. Since several other converter topologies have operation principles that are similar to those of the three-phase voltage source inverter, the simulation of the predictive control algorithm in this converter can serve as a starting point for further developments.

Figure A.1 shows the MATLAB/Simulink model used for simulation of the predictive current control of the voltage source inverter described in Chapter 4. The simulation diagram is composed of five major elements: the references, coordinate transformations, predictive control algorithm, inverter model, and load model.

The three-phase current references are generated by sine wave sources (block 1), which are configured with the desired peak amplitude, frequency, and phase angle. The predictive control algorithm can be directly implemented for three-phase currents. However, in order to reduce the number of predictions, the control can be performed in two-phase complex

Predictive Control of Power Converters and Electrical Drives, First Edition. Jose Rodriguez and Patricio Cortes.

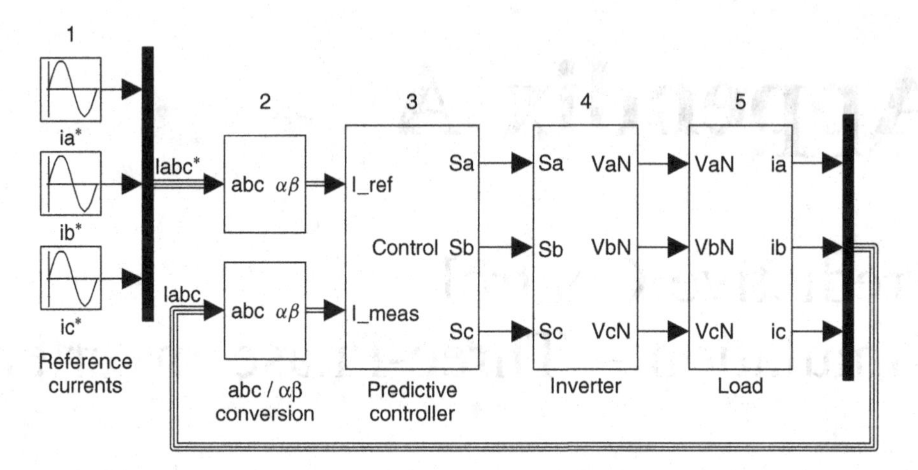

Figure A.1 Simulink model for simulation of predictive current control of a voltage source inverter

coordinates ($\alpha\beta$ coordinates). Since both the reference current and the load current measurements are three-phase variables, the coordinate transformation needs to be applied to each signal. In some applications where the reference current is already in $\alpha\beta$ coordinates, this step is not required. The transformation from *abc* to $\alpha\beta$ coordinates can be accomplished through (4.17), which can be separated into its real and imaginary components by

$$i_\alpha = \frac{2}{3}\left(i_a - \frac{1}{2}i_b - \frac{1}{2}i_c\right) \tag{A.1}$$

$$i_\beta = \frac{2}{3}\left(\frac{\sqrt{3}}{2}i_b - \frac{\sqrt{3}}{2}i_c\right) \tag{A.2}$$

These transformation equations are implemented in block 2 of the predictive control diagram of Figure A.1, whose detailed layout is shown in Figure A.2.

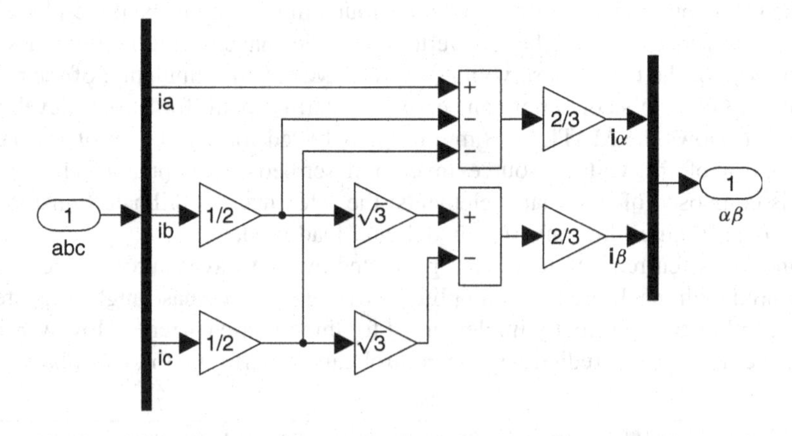

Figure A.2 Transformation from *abc* to $\alpha\beta$ coordinates

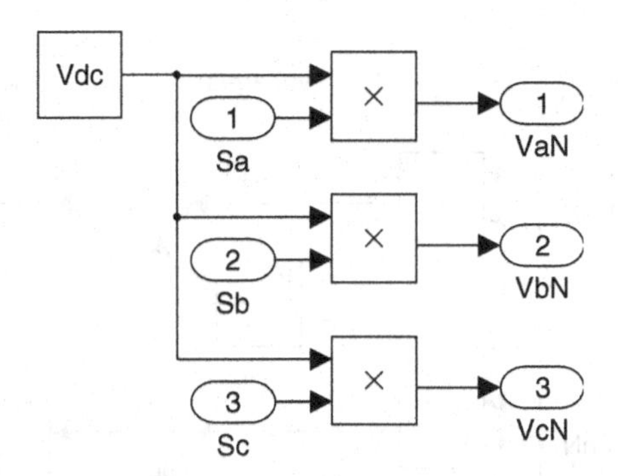

Figure A.3 Simulink model for the three-phase voltage source inverter

The core of the predictive control algorithm is implemented in an embedded MATLAB function (block 3) whose inputs are the reference and measured currents expressed in $\alpha\beta$ coordinates. The block outputs are the gating signals to be applied to the inverter. The MATLAB code for the predictive algorithm will be explained in detail in Section A.1.2.

The inverter is modeled as shown in Figure A.3, where the voltage of each inverter leg with respect to the negative busbar (N) is calculated by multiplying the DC link voltage by the corresponding gating signal, according to (4.5)–(4.7). In this model the DC link is assumed to be an ideal DC source. The multiplication of the gating signal by the DC link voltage inherently implies that the power semiconductors are modeled as ideal switches.

The load model for the simulation is shown in Figure A.4. This model is obtained for the variables defined according to the topological description shown in Figure 4.2. From Kirchhoff's voltage law, the load voltages for each phase are given by

$$v_{an} = v_{aN} - v_{nN} \tag{A.3}$$

$$v_{bn} = v_{bN} - v_{nN} \tag{A.4}$$

$$v_{cn} = v_{cN} - v_{nN} \tag{A.5}$$

The common-mode voltage v_{nN} can be determined by adding (4.13)–(4.15), which yields

$$\begin{aligned} & v_{aN} + v_{bN} + v_{cN} \\ & = L\frac{d}{dt}(i_a + i_b + i_c) + R(i_a + i_b + i_c) + (e_a + e_b + e_c) + 3v_{nN} \end{aligned} \tag{A.6}$$

Considering the star connection of the load, Kirchhoff's current law states that $i_a + i_b + i_c = 0$ in (A.6). Furthermore, assuming that the load back-emf is a balanced three-phase voltage, $e_a + e_b + e_c$ is also zero. Replacing both conditions in (A.6) and solving for v_{nN} gives the following expression for the common-mode voltage:

$$v_{nN} = \frac{1}{3}(v_{aN} + v_{bN} + v_{cN}) \tag{A.7}$$

The dynamics of the resistive–inductive load are represented through linear continuous-time transfer functions, which are obtained by substituting (A.3)–(A.5) into

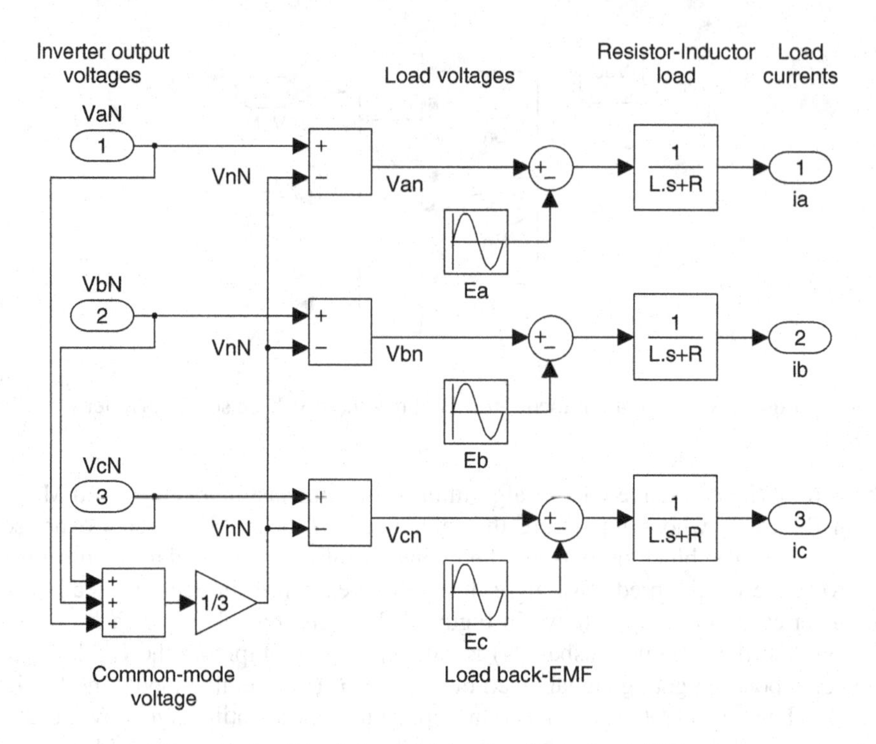

Figure A.4 Simulink model for the three-phase *RL* load

(4.13)–(4.15) to yield

$$v_{an} = L\frac{di_a}{dt} + Ri_a + e_a \tag{A.8}$$

$$v_{bn} = L\frac{di_b}{dt} + Ri_b + e_b \tag{A.9}$$

$$v_{cn} = L\frac{di_c}{dt} + Ri_c + e_c \tag{A.10}$$

By applying the Laplace transform to (A.8)–(A.10), the transfer functions from voltage to current at the *RL* load are obtained:

$$\frac{I_a}{V_{an} - E_a} = \frac{1}{Ls + R} \tag{A.11}$$

$$\frac{I_b}{V_{bn} - E_b} = \frac{1}{Ls + R} \tag{A.12}$$

$$\frac{I_c}{V_{cn} - E_c} = \frac{1}{Ls + R} \tag{A.13}$$

where the upper case variables represent the Laplace transforms of the respective lower case variables in the time domain. The output of the transfer functions is the load current of each phase, while their input is a voltage obtained by subtracting the back-emf from the corresponding load voltage. The back-emf is considered to be sinusoidal with constant amplitude and frequency, and is simulated by standard sine wave blocks.

A.1.1 Definition of Simulation Parameters

The values of all the parameters required by the model blocks can be defined directly in each block, initialized in the work space of MATLAB or contained in a m-file that needs to be executed before starting the simulation. The latter allows one to easily initialize and edit the simulation parameters, and reduces the risk of mismatches between the different blocks of the model. The code for the m-file with the parameter definition is presented in Code A.1. The sampling time (`Ts`) of the predictive algorithm is configured in line 4. Lines 6 to 9 in the code are used to set the load parameters, namely resistance (`R`), inductance (`L`), back-emf peak amplitude (`e`), and back-emf frequency (`f_e`); the DC link voltage (`Vdc`) is defined in line 10. The reference current amplitude (`I_ref_peak`) and frequency (`f_ref`) are defined in lines 12 and 13, respectively. For both the reference current and the back-emf the frequency is given in rad/s, as required by the Simulink sine wave blocks used for generating those signals.

The eight available voltage vectors of the inverter are also defined in this m-file (lines 15 to 23), according to Table 4.2. The corresponding switching states for each vector are also defined (line 25), so that the predictive algorithm is able to select one of these states for the controller output.

Code A.1 Parameters for the predictive control simulation.

```
 1 % Variables required in the control algorithm
 2 global Ts R L v states
 3 % Sampling time of the predictive algorithm [s]
 4 Ts = 25e-6;
 5 % Load parameters
 6 R = 10;              % Resistance [Ohm]
 7 L = 10e-3;           % Inductance [H]
 8 e = 100;               % Back-EMF peak amplitude [V]
 9 f_e = 50*(2*pi);         % Back-EMF frequency [rad/s]
10 Vdc = 520;           % DC-link voltage [V]
11 % Current reference
12 I_ref_peak = 10;          % Peak amplitude [A]
13 f_ref = 50*(2*pi);         % Frequency [rad/s]
14 % Voltage vectors
15 v0 = 0;
16 v1 = 2/3*Vdc;
17 v2 = 1/3*Vdc + 1j*sqrt(3)/3*Vdc;
18 v3 = -1/3*Vdc + 1j*sqrt(3)/3*Vdc;
19 v4 = -2/3*Vdc;
20 v5 =-1/3*Vdc - 1j*sqrt(3)/3*Vdc;
21 v6 = 1/3*Vdc - 1j*sqrt(3)/3*Vdc;
22 v7 = 0;
23 v = [v0 v1 v2 v3 v4 v5 v6 v7];
24 % Switching states
25 states = [0 0 0;1 0 0;1 1 0;0 1 0;0 1 1;0 0 1;1 0 1;1 1 1];
```

In addition to the parameters of the block, Simulink also requires definition of the simulation parameters (such as solver type, start/stop time, etc.). The most important

Table A.1 Suggested simulation configuration parameters

Parameter	Value
Start time	0.0 [s]
Stop time	0.15 [s]
Solver type	Fixed step
Solver	ode 5 (Dormand–Prince)
Fixed-step size	1e-6 [s]
Tasking mode	Singletasking

configuration parameters of the simulation are presented in Table A.1. The remaining configuration parameters are set by default.

A.1.2 MATLAB® Code for Predictive Current Control

The predictive algorithm was implemented using an embedded MATLAB function block (block 3 in Figure A.1). This block must be configured to operate with a discrete update method at the sample time defined for the predictive algorithm (`Ts`, as explained in Section A.1.1). This configuration can be easily done using the Simulink model explorer, which also allows the variables of the algorithm to be defined as inputs, outputs, or parameters.

The algorithm for the predictive current control method, as explained in Section 4.8, is presented in Code A.2. The first line of the code declares the control function, where the outputs are the three gating signals `Sa`, `Sb`, and `Sc`. The inputs of the algorithm are the reference current (`I_ref`) and the measured current (`I_meas`), both expressed in $\alpha\beta$ coordinates. The remaining arguments of the function are parameters required for current prediction and output state selection.

The algorithm in this example requires two variables to be recalled from the previous sampling instant: the optimum vector selected by the algorithm (`x_old`) and the instant current measurement (`i_old`). These variables are used to estimate the load back-emf, given by (4.23) in Section 4.6. Lines 5 to 8 are used to declare these persistent variables and initialize their values, while the back-emf (`e`) is estimated in line 15.

The reference current (`ik_ref`) and the measured current (`ik`) at sampling instant k are accessed in lines 11 and 13 of the code, respectively; the value of the current at instant $k-1$ is updated in line 17.

The optimization procedure is performed by the code segment between lines 18 and 29. The procedure sequentially selects one of the eight possible voltage vectors (line 20) and applies it to the equation in line 22 in order to obtain the current prediction at instant $k+1$ (`ik1`). The cost function in line 24 is used to evaluate the error between the reference and the predicted current in the next sampling instant for each voltage vector. The optimal values of the cost function (`g_opt`) and the optimum voltage vector (`x_opt`) are selected using the code in lines 26 to 29. The value of (`g_opt`) is initialized with a very high number in line 9. The value of the previous optimum voltage vector is updated in line 32. Finally, the output switching states corresponding to the optimum voltage vector are generated between lines 34 and 36.

Code A.2 Predictive control algorithm.

```
 1 function [Sa,Sb,Sc] = control(I_ref,I_meas)
 2 % Variables defined in the parameters file
 3 global Ts R L v states
 4 % Optimum vector and measured current at instant k-1
 5 persistent x_old i_old
 6 % Initialize values
 7 if isempty(x_old), x_old = 1; end
 8 if isempty(i_old), i_old = 0+1j*0; end
 9 g_opt = 1e10;
10 % Read current reference inputs at sampling instant k
11 ik_ref = I_ref(1) + 1j*I_ref(2);
12 % Read current measurements at sampling instant k
13 ik = I_meas(1) + 1j*I_meas(2);
14 % Back-EMF estimate
15 e = v(x_old) - L/Ts*ik - (R - L/Ts)*i_old;
16 % Store the measured current for the next iteration
17 i_old = ik;
18 for i = 1:8
19     % i-th voltage vector for current prediction
20     v_o1 = v(i);
21     % Current prediction at instant k+1
22     ik1 = (1 - R*Ts/L)*ik + Ts/L*(v_o1 - e);
23     % Cost function
24     g = abs(real(ik_ref - ik1)) + abs(imag(ik_ref - ik1));
25     % Selection of the optimal value
26     if (g<g_opt)
27       g_opt = g;
28       x_opt = i;
29     end
30 end
31 % Store the present value of x_opt
32 x_old = x_opt;
33 % Output switching states
34 Sa = states(x_opt,1);
35 Sb = states(x_opt,2);
36 Sc = states(x_opt,3);
```

Appendix B

Predictive Control Simulation – Torque Control of an Induction Machine Fed by a Two-Level Voltage Source Inverter

Figure B.1 shows the MATLAB®/Simulink® model used for simulation of the predictive torque control (PTC) of an induction machine fed by a two-level voltage source inverter described in Chapter 8. The simulation diagram contains six main elements: reference speed generation, PI speed controller, predictive control algorithm, inverter model, coordinate transformation, and induction machine model.

The reference speed (block 1) in the simulation layout can be a constant, a step block, or any other signal according to the simulation needs. For the reference speed tracking, a PI controller (block 2) is used. The PI controller receives the error signal and computes the torque reference for the predictive controller. Figure B.2 shows a discrete-time implementation of the PI controller. There are two tuning parameters in this controller: the proportional gain Kp and the integral gain Ki. For the design of these parameters, the transfer function between the rotor speed and the electrical torque is obtained from (8.12) through application of the Laplace transform:

$$\frac{\omega_m(s)}{T(s)} = \frac{1}{Js} \tag{B.1}$$

Starting from (B.1), a variety of well-known methods for the design of PI controllers can be applied. The values of the controller parameters employed in this simulation will be presented in Section B.1.

A saturation block is included at the output of the PI controller in order to keep the amplitude of the torque reference within the limits of the simulated machine. The presence of this constraint can give rise to the windup phenomenon, which occurs when the integrator in the controller continues to integrate while the input is limited, and has a detrimental effect on the transient response. A simple anti-windup scheme implemented in

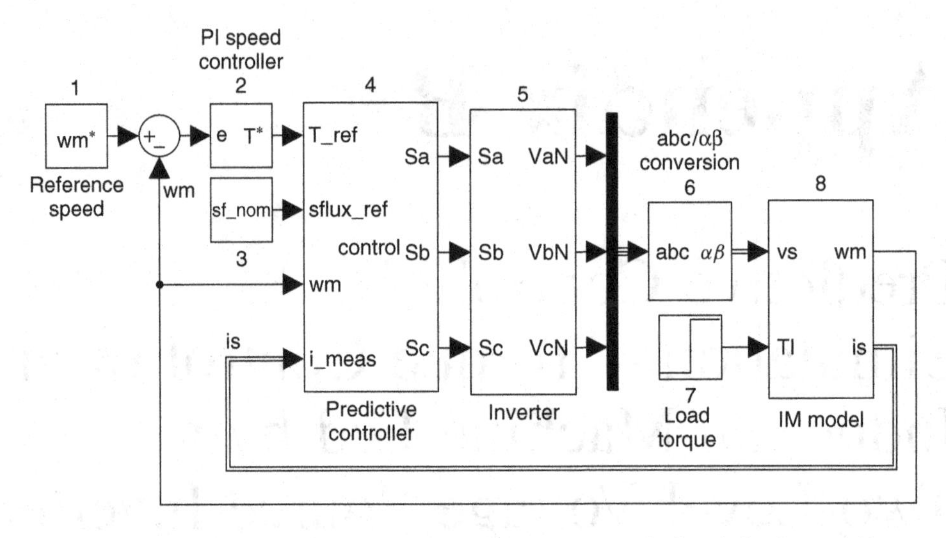

Figure B.1 Simulink model for the predictive torque control of an induction machine

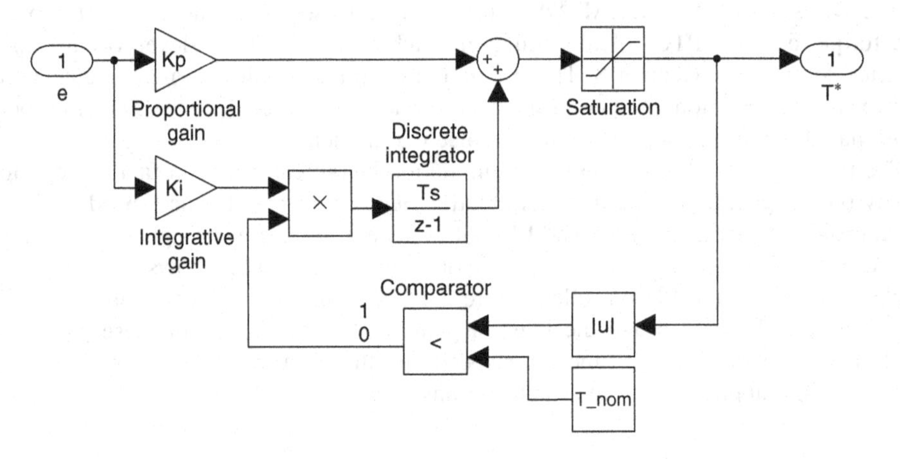

Figure B.2 PI speed controller with anti-windup

this simulation operates by comparing the absolute value of the output reference torque to the limit imposed on this variable. The output of the comparator can take values of 1 or 0 and multiplies the input signal to the integrator. In case the magnitude of the reference torque is under the limit, the output of the comparator is 1 and the integrator operates normally. On the other hand, when the output is saturated the input to the integrator is zero, thus preventing the windup effect.

The reference stator flux magnitude corresponds to the second input of the predictive controller, and is a constant generated by block 3 in Figure B.1. The predictive algorithm is implemented in an embedded MATLAB function (block 4). This block also has the measured rotor speed and stator currents as inputs, while its outputs are the gating signals to be applied to the inverter. The MATLAB code for the predictive algorithm will be explained in detail in Section B.2.

The two-level inverter (block 5) is modeled as explained in Section A.1 and is depicted in Figure A.3. Since the induction machine model used in this simulation is developed in the $\alpha\beta$ frame, the output voltages of the inverter are processed by a coordinate conversion block identical to that explained in Section A.1.

The dynamic model of the induction machine used in the simulation (block 8) is shown in Figure B.3. This block has the stator voltage and the load torque as inputs, while its outputs are the stator current and the rotor speed. It is possible to distinguish three main parts in the model layout. In the first one, the α and β components of the stator voltage are merged into a complex signal, which allows direct implementation of the complex model explained in Section 8.2. The second section corresponds to the stator and rotor dynamics, expressed by (8.14) and (8.15). There are two integrations of complex signals in the model, one for each of the state variables (stator current and rotor flux). Since the integrators in Simulink are not able to handle complex signals, the integration is implemented separately for the real and imaginary components, as shown in Figure B.4. The gain blocks in the model contain both the notation used for variables in Section 8.2 (inside labels) and the notation used in the MATLAB code in Section B.1 (outside labels).

The third section of the model contains the calculation of the electromagnetic torque and the mechanical subsystem of the machine. The electromagnetic torque can be expressed as a function of two state variables of the model. For example, (8.11) presents two possibilities: stator flux and stator current; and rotor flux and rotor current. For the sake of simplicity, in this simulation the torque is calculated in terms of the same state variables used for the rest of the model, that is, rotor flux and stator current:

$$T = \frac{3}{2} p Im\{\bar{\psi}_r \cdot \mathbf{i}_s\}. \tag{B.2}$$

The dynamics of the rotor speed are given by (8.12), where the inputs are the electromagnetic torque T and the load torque disturbance T_l; the output is the rotor speed of the machine ω_m. As expressed by (8.13), the rotor speed needs to be multiplied by the number of pole pairs p in order to obtain the electric rotor speed w.

B.1 Definition of Predictive Torque Control Simulation Parameters

The values of all the parameters required by the model blocks of the simulation are written in a m-file, whose contents are presented in Code B.1. The sampling time for the PTC is defined in line 4 of the code. The parameters for the discrete PI speed controller are given between lines 6 and 8. Lines 10 to 18 are used to define the machine parameters. The nominal stator flux (`sf_nom`) in line 17 is used as the stator flux reference for the predictive controller (block 3 in Figure B.1), whereas the nominal torque (`T_nom`) in line 18 fixes the limits for the saturator inside the PI controller. The DC link voltage of the inverter is set in line 20. A number of auxiliary constants introduced in (8.14) and (8.15) are defined between lines 22 and 27. The equivalents of these variables to those defined in Section 8.2 are the following: `ts`$=\tau_s$, `tr`$=\tau_r$, `sigma`$=\sigma$, `kr`$=k_r$, `r_sigma`$=R_\sigma$, `t_sigma`$=t_\sigma$.

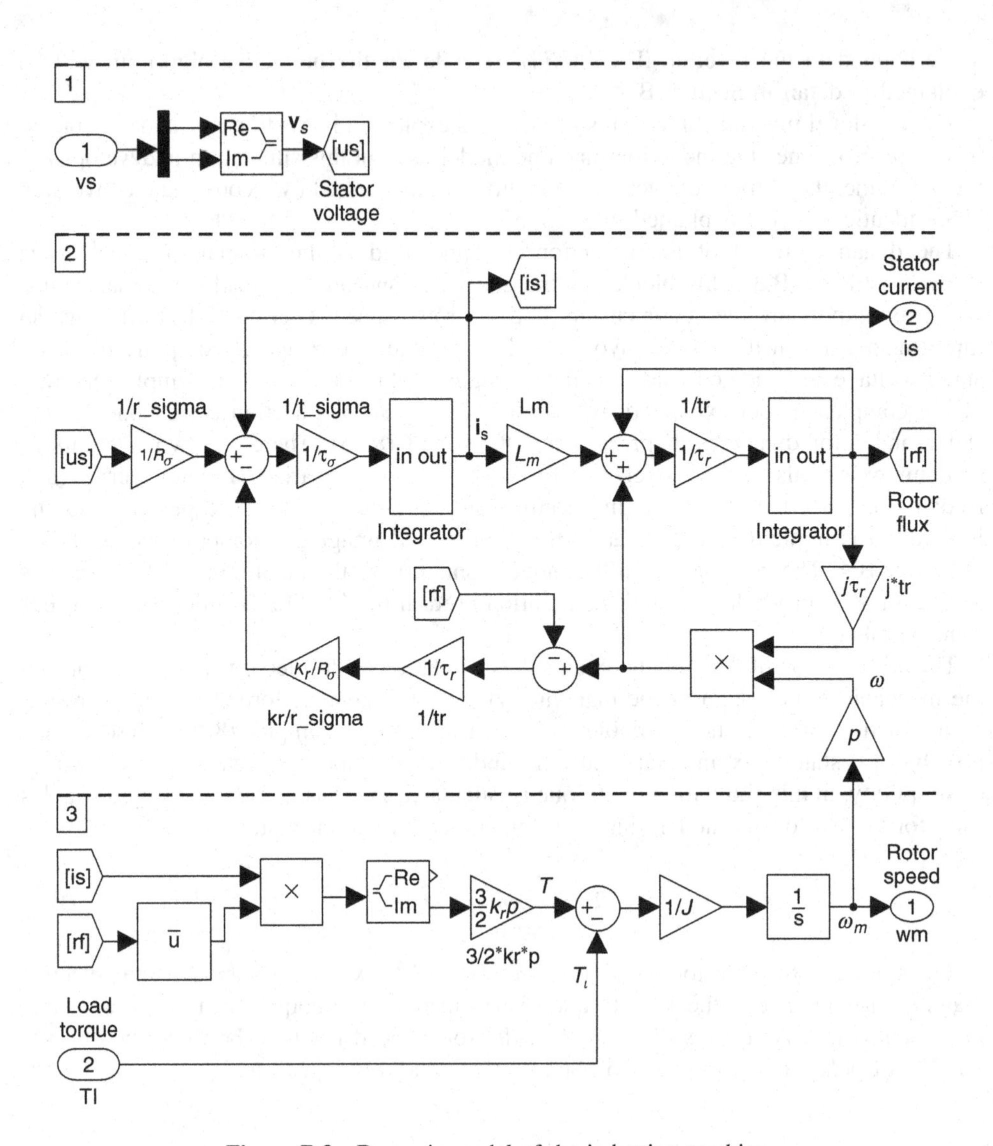

Figure B.3 Dynamic model of the induction machine

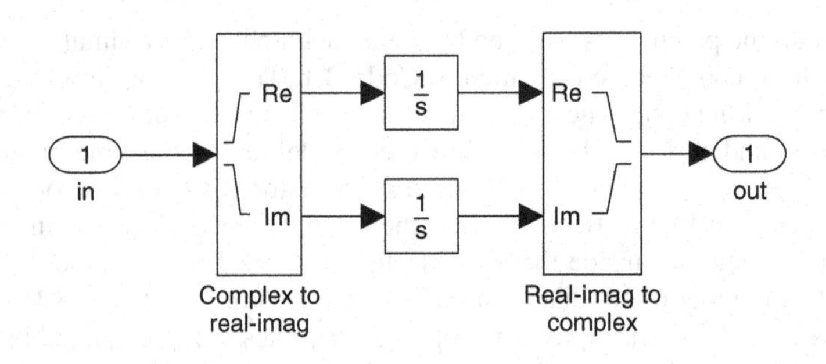

Figure B.4 Complex integrator implementation

Table B.1 Suggested simulation configuration parameters

Parameter	Value
Start time	0.0 [s]
Stop time	3.0 [s]
Solver type	Fixed step
Solver	ode 5 (Dormand–Prince)
Fixed-step size	1e-5 [s]
Tasking mode	Singletasking

The weighting parameter `lambda` of the PTC is calculated in line 29 according to (8.28). Finally, the voltage vectors of the inverter and the corresponding switching states are defined between lines 31 and 41.

The most important configuration parameters of the simulation are presented in Table B.1. The remaining configuration parameters are set by default.

B.2 MATLAB® Code for the Predictive Torque Control Simulation

The predictive controller was implemented using an embedded MATLAB function block (block 4 in Figure B.1). It is important that this block is configured to operate at the sampling time defined for the predictive algorithm in the previous section. Also, some of the arguments of the control function should be defined as inputs while the rest correspond to parameters. Both configurations can be easily done using the Simulink model explorer.

The control algorithm, which was explained in Section 8.4, is presented in Code B.2. The first line in the code declares the control function, where the outputs are the gating signals `Sa`, `Sb`, and `Sc`. The inputs of the algorithm are the torque and flux references (`T_ref` and `sflux_ref`) and the measured rotor speed and stator current (`wm` and `i_meas`). The remaining arguments of the function correspond to parameters required for the predictions and output state selection.

The PTC algorithm requires two variables to be recalled from the previous sampling instant: the optimum voltage vector selected by the algorithm (`x_opt`) and the stator flux estimate (`Fs`). These variables are used to estimate the stator flux, according to (8.18). Lines 4 to 6 are used to declare these persistent variables and to initialize their values, whereas the stator flux (`Fs`) is estimated in line 8. The rotor flux is also estimated, according to (8.22), in line 10.

The prediction–optimization procedure is performed with the code segment between lines 12 and 31. The algorithm sequentially selects one of the eight available voltage vectors (line 14) and applies it to the equations in lines 16 and 18 in order to obtain the stator flux and current predictions at instant $k + 1$ (`Fsp1` and `Isp1`, respectively). The electromagnetic torque is also predicted in line 21. Then, the cost function in line 23 is used to evaluate the error between the reference and predicted torque and stator flux in the next sampling instant for each of the voltage vectors. The weighting factor `lambda` is employed to adjust the relative importance of the torque versus the flux control. The optimum value of the cost function (`g_opt`), whose initial value is assigned in line 11, is stored in lines 24 to 27. The optimum voltage vector (`x_opt`) is also selected in this part of the algorithm. Finally, the output switching states that correspond to the optimum voltage vector are sent to the outputs between lines 33 and 35.

Code B.1 Parameters for the PTC.

```
 1 % Variables required by the control algorithm
 2 global Ts Rs Lr Lm Ls p tr kr r_sigma t_sigma lambda v states
 3 % Sampling time of the predictive algorithm [s]
 4 Ts = 4e-5;
 5 % PI speed controller parameters
 6 Tsw = 0.002;  % Sampling time of the PI controller [s]
 7 Kp = 3.016;  % Proportional gain
 8 Ki = 0.141;  % Integrative gain
 9 % Machine parameters
10 J = 0.062; % Moment of inertia [kg m^2]
11 p = 1;     % Pole pairs
12 Lm = 170e-3;     % Magnetizing inductance [H]
13 Ls = 175e-3;     % Stator inductance [H]
14 Lr = 175e-3;     % Rotor inductance [H]
15 Rs = 1.2;        % Stator resistance [Ohm]
16 Rr = 1;          % Rotor resistance [Ohm]
17 sf_nom = 0.71;   % Nominal stator flux [Wb]
18 T_nom = 20;      % Nominal torque [Nm]
19 % DC-link voltage [V]
20 Vdc = 520;
21 % Auxiliary constants
22 ts = Ls/Rs;
23 tr = Lr/Rr;
24 sigma = 1-(((Lm)^2)/(Lr*Ls));
25 kr = Lm/Lr;
26 r_sigma = Rs+kr^2*Rr;
27 t_sigma = sigma*Ls/r_sigma;
28 % Weighting factor for the cost function of PTC
29 lambda = T_nom/sf_nom;
30 % Voltage vectors
31 v0 = 0;
32 v1 = 2/3*Vdc;
33 v2 = 1/3*Vdc + 1j*sqrt(3)/3*Vdc;
34 v3 = -1/3*Vdc + 1j*sqrt(3)/3*Vdc;
35 v4 = -2/3*Vdc;
36 v5 = -1/3*Vdc - 1j*sqrt(3)/3*Vdc;
37 v6 = 1/3*Vdc - 1j*sqrt(3)/3*Vdc;
38 v7 = 0;
39 v = [v0 v1 v2 v3 v4 v5 v6 v7];
40 % Switching states
41 states = [0 0 0;1 0 0;1 1 0;0 1 0;0 1 1;0 0 1;1 0 1;1 1 1];
```

Code B.2 PTC algorithm.

```
 1 function [Sa,Sb,Sc] = control(T_ref,sflux_ref,wm,i_meas)
 2 % Variables defined in the parameters file
 3 global Ts Rs Lr Lm Ls p tr kr r_sigma t_sigma lambda v states
 4 persistent x_opt Fs
 5 if isempty(x_opt), x_opt = 1; end
 6 if isempty(Fs), Fs = 0 + 0i*1; end
 7 % Stator flux estimate
 8 Fs = Fs + Ts*(v(x_opt) - Rs*i_meas);
 9 % Rotor flux estimate
10 Fr = Lr/Lm*Fs+i_meas*(Lm-Lr*Ls/Lm);
11 g_opt = 1e10;
12 for i = 1:8
13     % i-th voltage vector for current prediction
14     v_o1 = v(i);
15     % Stator flux prediction at instant k+1
16     Fsp1 = Fs + Ts*v_o1 - Rs*Ts*i_m_eas;
17     % Stator current prediction at instant k+1
18     Isp1 = (1+Ts/t_sigma)*i_meas+Ts/(t_sigma+Ts)*...
19         (1/r_sigma*((kr/tr-kr*1i*wm)*Fr+v_o1));
20     % Torque prediction at instant k+1
21     Tp1 = 3/2*p*imag(conj(Fsp1)*Isp1);
22     % Cost function
23     g = abs(T_ref - Tp1)+ lambda*abs(sflux_ref-abs(Fsp1));
24     if (g<g_opt)
25     g_opt = g;
26     x_opt = i;
27     end
28 end
29 %***************************************
30 % Optimization
31  [~, x_opt] = min(g);
32 % Output switching states
33 Sa = states(x_opt,1);
34 Sb = states(x_opt,2);
35 Sc = states(x_opt,3);
```

Appendix C

Predictive Control Simulation – Matrix Converter

C.1 Predictive Current Control of a Direct Matrix Converter

As mentioned in Chapter 7, the matrix converter (MC) consists of an array of bidirectional switches, which are used to directly connect the power supply to the load without using any DC-link or large energy storage elements. Also, this is a more complex topology with respect to the three-phase voltage source inverter; throughout its simulation and predictive control algorithm implementation it will be demonstrated that this method can be applied easily to this converter.

Figure C.1 shows the MATLAB®/Simulink® model used for simulation of the predictive current control of the direct MC described in Chapter 7. The simulation diagram is composed of six major elements: the references, predictive control algorithm, and AC-supply, input filter, converter, and load models. The three-phase output current references are generated by sine wave sources (block 1), which are configured with the desired peak amplitude, frequency, and phase angle.

The core of the predictive control algorithm is implemented in an embedded MATLAB function (block 2) whose inputs are the grid voltages and currents, capacitor voltages, measured load currents, and reference load currents. The reference of the instantaneous reactive power is set to zero in order to have zero instantaneous reactive power on the input side of the system. The block outputs are the gating signals to be applied to the direct MC. The MATLAB code for the predictive algorithm will be explained in detail in Section C.1.2.

The AC-supply is generated by sine wave sources (block 3), which are configured with the desired peak amplitude, frequency, and phase angle.

As mentioned in Chapter 7, an $L_f C_f$ filter is required at the input of the converter to reduce the high-frequency current harmonics caused by the switching operation and the inductive nature of the AC-line. The filter model is indicated in block 4 and consists of a second-order system described by (7.25) and (7.26). The inputs of this block are the

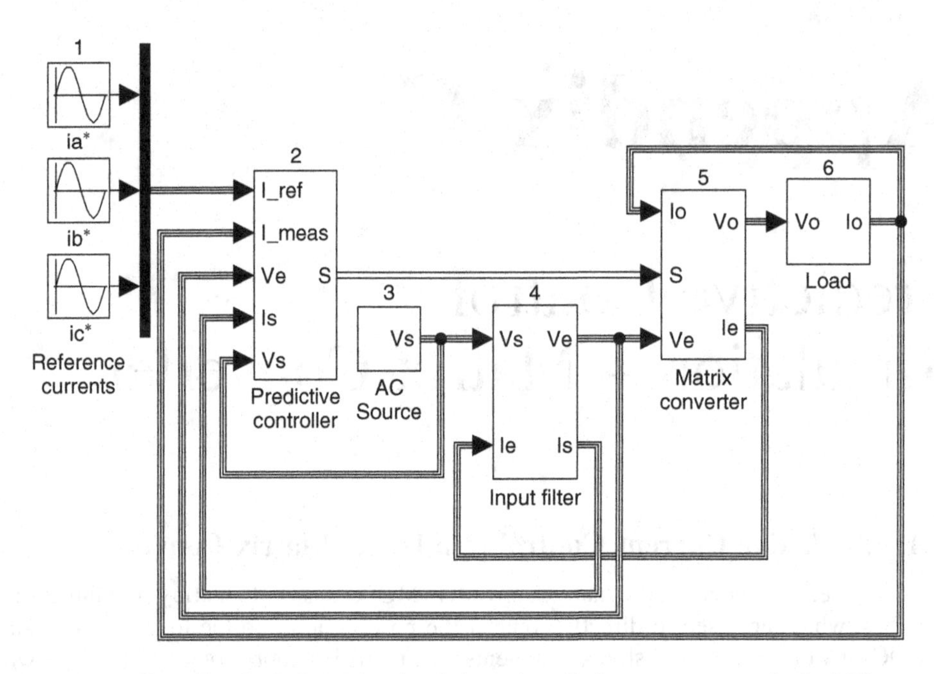

Figure C.1 Simulink model for simulation of predictive current control of a MC

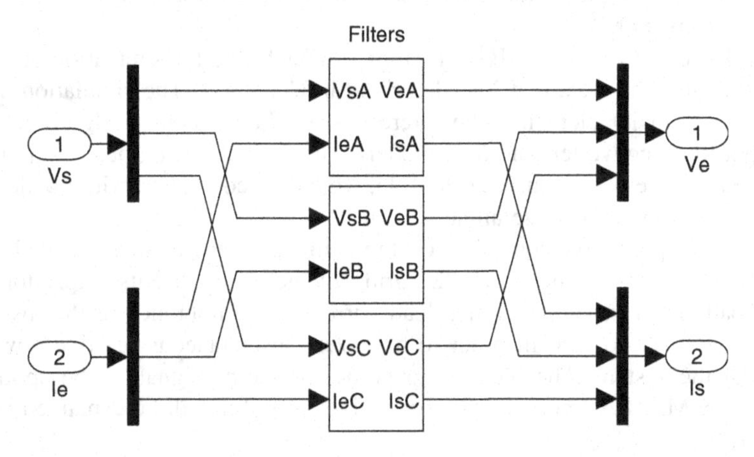

Figure C.2 Simulink model of the input filter for simulation of predictive current control in a MC

grid voltage and the converter currents, and the outputs are the capacitor voltages and grid current as indicated in Figure C.2, where each block represents the filter stage of one phase. The filter is implemented using the corresponding transfer functions, as detailed in Figure C.3.

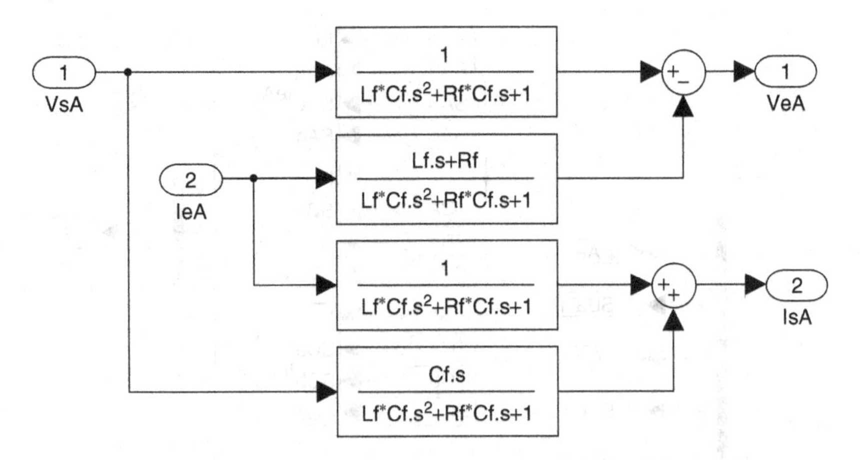

Figure C.3 Input filter model used in the simulation

The MC is modeled as shown in Figure C.4 and Figure C.5, where the phase-to-neutral voltage of each output leg is calculated by multiplying the input voltage by the corresponding gating signal, according to (7.2). Similarly, the input current vector is given by the load currents and the corresponding gating signal, as indicated in (7.5).

The load model for the simulation is the same as the model shown previously. This model is obtained for the variables defined according to the topological description shown in Figure 7.1. The dynamics of the resistive–inductive load are represented through linear continuous-time transfer functions, which are obtained according to (7.22).

C.1.1 Definition of Simulation Parameters

The code for the m-file with the definition of parameters is presented in Code C.1. The sampling time (`Ts`) of the predictive algorithm is configured in line 4. The source voltage (`Vs`) and frequency (`fs`) are set in lines 6 and 8, respectively. Lines 10 to 12 in the code are used to set the input filter parameters, namely, inductance (`Lf`), resistance (`Rf`), and capacitance (`Cf`). Lines 14 to 15 are used to set the load parameters, namely, resistance (`R`), inductance (`L`). The reference current amplitude (`i_ref`) and frequency (`w_ref`) are defined in lines 17 and 18, respectively. The frequency of the reference current is given in rad/s, as required by the Simulink sine wave blocks used for generating these signals. In (7.36), the weighting factor A handles the relevance of each objective and this value is set in line 20. The state space model of the input filter is written in lines 22 to 26. By discretizing the filter model, using the MATLAB function *c2d*, the values from 28 to 36 are obtained, which correspond to the constants of (7.31). The 27 available switching states for each vector are also defined (lines 38 to 64), so that the predictive algorithm is able to select one of these states for the controller output.

In addition to the parameters of the block, MATLAB/Simulink also requires definitions for the simulation parameters (such as solver type, start/stop time, etc.). The suggested

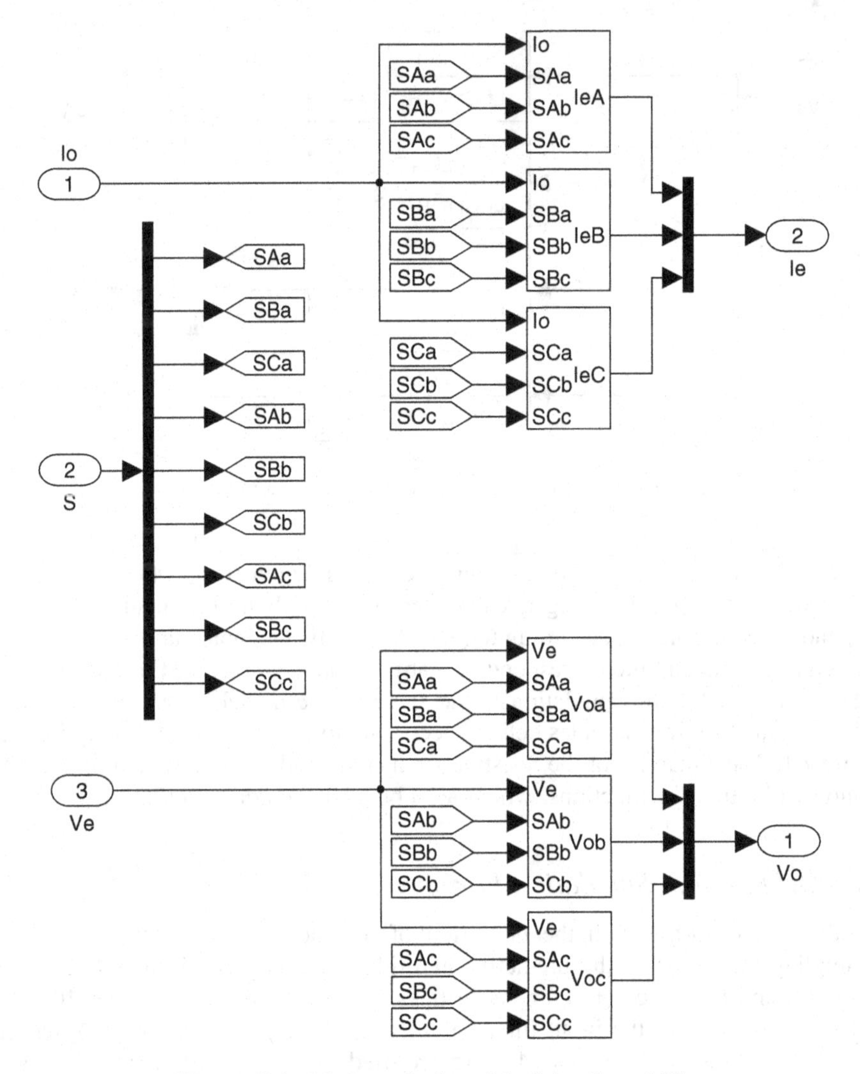

Figure C.4 Mathematical model of the direct MC

configuration parameters of the simulation are presented in Table C.1. The remaining configuration parameters are set by default.

C.1.2 MATLAB® Code for Predictive Current Control with Instantaneous Reactive Power Minimization

The predictive algorithm is implemented using an embedded MATLAB function block (block 2 in Figure C.1). The sampling time T_s for the control algorithm is set in the embedded MATLAB function block properties. In this example two objectives are accomplished: first, the output current presents a good tracking to its reference and, second, the instantaneous reactive power on the input side is minimized.

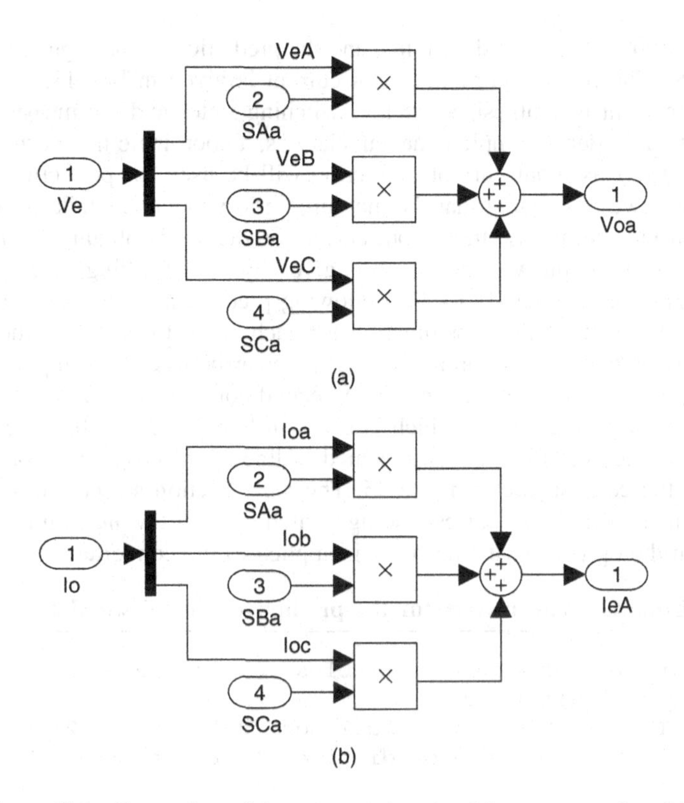

Figure C.5 Generation of the output voltage and input current in a MC

Table C.1 Simulation configuration parameters

Parameter	Value
Start time	0.0 [s]
Stop time	0.15 [s]
Solver type	Fixed step
Solver	ode 5 (Dormand–Prince)
Fixed-step size	1e-6 [s]

The algorithm for the predictive current control method, explained in Section 7.4.3, is presented in Code C.2. As mentioned previously, the inputs of the algorithm are grid voltages and currents, capacitor voltages, measured load currents, and reference load currents. The remaining arguments of the function are parameters required for the input and load current prediction and state selection. The most important section in this algorithm is given between lines 10 and 39, where the prediction of the load current and the instantaneous input reactive power is done for each possible switching state of the converter. In line 15 prediction of the input current is obtained, which is given by the measurement of the grid voltages, capacitor voltages, grid currents, and the prediction of the input current for each switching state. The prediction of the input current is given in line 13, where it can be seen that this current is synthesized by the switching states and the measurement of the output currents. In order to simplify the calculations, a coordinate transformation is made in lines 17 to 18. This input current prediction will be used for predicting the instantaneous reactive power, assuming that the measured grid voltage is similar to its prediction in the $k + 1$ instant. Similarly, the output current prediction is obtained in line 26 based on prediction of the output voltage which is given by the switching states and capacitor voltages as indicated in lines 21 to 24. Following prediction of the necessary variables, it is necessary to evaluate the cost function for each control objective. The first control objective is the output current control, where the error between the output current reference and the prediction is given in line 31. A second control objective is minimization of the instantaneous reactive power, which is given in line 32. Both objectives are merged in a single so-called cost function as indicated in line 33. The optimization procedure is performed by the code segment in line 35. The *min* function selects the minimal value of the cost function and the corresponding switching states which minimize this cost function. Then the optimal switching state is applied to the converter.

Code C.1 Parameters for the predictive control simulation.

```
 1 % Declaration of global variables required by the
       control algorithm
 2 global Ts R L A Aq11 Aq12 Aq21 Aq22 Bq11 Bq12 Bq21 Bq22 S
 3 % Sampling time of the predictive algorithm [s]
 4 Ts = 10e-6;
 5 % Source voltage [V]
 6 Vs = 180;
 7 % Source frequency [f]
 8 ws = 2*pi*50;
 9 % Input filter parameters
10 Lf = 400e-6;
11 Rf = 0.5;
12 Cf = 21e-6;
13 % Load parameters
14 R = 10;
15 L = 30e-3;
16 % References
17 i_ref = 8;
18 w_ref = 2*pi*30;
19 % Weighting factor instantaneous reactive power minimization
20 A = 0.008;
21 % Input filter model
22 Ai = [0 1/Cf; -1/Lf -Rf/Lf];
23 Bi = [0 -1/Cf; 1/Lf 0];
24 Ci = [1 0; 0 1];
25 Di = [0 0;0 0];
26 Gin = ss(Ai,Bi,Ci,Di);
27 % Discretization of the input filter model
28 Gq = c2d(Gin,Ts);
```

```
29 Aq11 = Gq.a(1,1);
30 Aq12 = Gq.a(1,2);
31 Aq21 = Gq.a(2,1);
32 Aq22 = Gq.a(2,2);
33 Bq11 = Gq.b(1,1);
34 Bq12 = Gq.b(1,2);
35 Bq21 = Gq.b(2,1);
36 Bq22 = Gq.b(2,2);
37 % Valid switching states of the Matrix Converter
38 S(:,:,1) = [1 0 0; 0 1 0; 0 0 1];
39 S(:,:,2) = [1 0 0; 0 0 1; 0 1 0];
40 S(:,:,3) = [0 1 0; 1 0 0; 0 0 1];
41 S(:,:,4) = [0 1 0; 0 0 1; 1 0 0];
42 S(:,:,5) = [0 0 1; 1 0 0; 0 1 0];
43 S(:,:,6) = [0 0 1; 0 1 0; 1 0 0];
44 S(:,:,7) = [1 0 0; 0 0 1; 0 0 1];
45 S(:,:,8) = [0 1 0; 0 0 1; 0 0 1];
46 S(:,:,9) = [0 1 0; 1 0 0; 1 0 0];
47 S(:,:,10) = [0 0 1; 1 0 0; 1 0 0];
48 S(:,:,11) = [0 0 1; 0 1 0; 0 1 0];
49 S(:,:,12) = [1 0 0; 0 1 0; 0 1 0];
50 S(:,:,13) = [0 0 1; 1 0 0; 0 0 1];
51 S(:,:,14) = [0 0 1; 0 1 0; 0 0 1];
52 S(:,:,15) = [1 0 0; 0 1 0; 1 0 0];
53 S(:,:,16) = [1 0 0; 0 0 1; 1 0 0];
54 S(:,:,17) = [0 1 0; 0 0 1; 0 1 0];
55 S(:,:,18) = [0 1 0; 1 0 0; 0 1 0];
56 S(:,:,19) = [0 0 1; 0 0 1; 1 0 0];
57 S(:,:,20) = [0 0 1; 0 0 1; 0 1 0];
58 S(:,:,21) = [1 0 0; 1 0 0; 0 1 0];
59 S(:,:,22) = [1 0 0; 1 0 0; 0 0 1];
60 S(:,:,23) = [0 1 0; 0 1 0; 0 0 1];
61 S(:,:,24) = [0 1 0; 0 1 0; 1 0 0];
62 S(:,:,25) = [1 0 0; 1 0 0; 1 0 0];
63 S(:,:,26) = [0 1 0; 0 1 0; 0 1 0];
64 S(:,:,27) = [0 0 1; 0 0 1; 0 0 1];
```

Code C.2 Predictive control algorithm.

```
1 function [Sopt] = MC_control(I_ref,Io,Ve,Is,Vs);
2 % Declaration of global variables required by the
      control algorithm
3 global Ts R L A Aq11 Aq12 Aq21 Aq22 Bq11 Bq12 Bq21 Bq22 S
4 % Output references in alpha-beta coordinates
5 Irefalpha = 2*(I_ref(1) - 0.5*I_ref(2) - 0.5*I_ref(3))/3;
6 Irefbeta = 2*(sqrt(3)*I_ref(2)*0.5 - sqrt(3)*I_ref(3)*0.5)/3;
7 % Initialization of the optimal value of the cost function
8 gopt = 1e10;
```

```
 9 % Calculation of predictions for the 27 switching states
10 for k = 1:27;
11  % input current vector is given by the switches state...
12  % and the load currents
13  Ie = S(:,:,k)'*Io;
14  % prediction of the source currents
15  Is_p_3f = Aq21*Ve + Aq22*Is + Bq21*Vs + Bq22*Ie;
16  % transformation to alpha-beta coordinates
17  Is_p_re = 2*(Is_p_3f(1) - 0.5*Is_p_3f(2)
          - 0.5*Is_p_3f(3))/3;
18  Is_p_im = 2*(sqrt(3)*Is_p_3f(2)*0.5
          - sqrt(3)*Is_p_3f(3)*0.5)/3;
19  % output voltage vector is given by the switches state...
20  % and the input voltage vector
21  VxN = S(:,:,k)*Ve;
22  Vo(1) = VxN(1) -(VxN(1) + VxN(2) + VxN(3))/3;
23  Vo(2) = VxN(2) -(VxN(1) + VxN(2) + VxN(3))/3;
24  Vo(3) = VxN(3) -(VxN(1) + VxN(2) + VxN(3))/3;
25  % prediction of the load currents
26  Io_p_3f = (1 - R*Ts/L)*Io + (Ts/L)*Vo;
27  % transformation to alpha-beta coordinates
28  Io_p_re = 2*(Io_p_3f(1) - 0.5*Io_p_3f(2)
          - 0.5*Io_p_3f(3))/3;
29  Io_p_im = 2*(sqrt(3)*Io_p_3f(2)*0.5
          - sqrt(3)*Io_p_3f(3)*0.5)/3;
30  % cost function calculation
31  g1 = (abs(Irefalpha - Io_p_re) + abs(Irefbeta-Io_p_im));
32  g2 = abs(Vs_p_re*Is_p_im - Vs_p_im*Is_p_re);
33  g = g1 + A*g2;
34  % optimization
35  if (g<gopt)
36  gopt = g;
37  eopt = k;
38  end
39 end
40
41 % Output switching states
41 SAa = S(1,1,eopt);
42 SBa = S(1,2,eopt);
43 SCa = S(1,3,eopt);
44 SAb = S(2,1,eopt);
45 SBb = S(2,2,eopt);
46 SCb = S(2,3,eopt);
47 SAc = S(3,1,eopt);
48 SBc = S(3,2,eopt);
49 SCc = S(3,3,eopt);
50 Sopt = [SAa SBa SCa SAb SBb SCb SAc SBc SCc;
```

Index

Predictive Control of Power Converters and Electrical Drives, First Edition. Jose Rodriguez and Patricio Cortes.